Gamma Cameras for Interventional and Intraoperative Imaging

Series in Medical Physics and Biomedical Engineering

Series Editors: John G Webster, E Russell Ritenour, Slavik Tabakov, and Kwan-Hoong Ng

Series in Medical Physics and Biomedical Engineering

Gamma Cameras for Interventional and Intraoperative Imaging

Edited by

Alan Perkins

University of Nottingham, United Kingdom

John E. Lees

University of Leicester, United Kingdom

CRC Press is an imprint of the
Taylor & Francis Group, an **informa** business

CRC Press
Taylor & Francis Group
6000 Broken Sound Parkway NW, Suite 300
Boca Raton, FL 33487-2742

First issued in paperback 2019

ISBN-13: 978-0-4987-2928-4 (hbk)
ISBN-13: 978-0-367-87350-9 (pbk)

Library of Congress Cataloging-in-Publication Data

Names: Perkins, A. C., editor. | Lees, John E., editor.
Title: Gamma cameras for interventional and intraoperative imaging / edited by Alan C. Perkins and John E. Lees.
Other titles: Series in medical physics and biomedical engineering.
Description: Boca Raton, FL : CRC Press, Taylor & Francis Group, [2016] | Series: Series in medical physics and biomedical engineering | Includes bibliographical references.
Identifiers: LCCN 2016022758| ISBN 9781498729284 (hardback ; alk. paper) | ISBN 1498729282 (hardback ; alk. paper) | ISBN 9781498729291 (e-book) | ISBN 1498729290 (e-book)
Subjects: LCSH: Scintillation cameras. | Nuclear medicine--Instruments. | Imaging systems in medicine. | Diagnostic imaging. | Point-of-care testing.
Classification: LCC RC78.5 .G36 2016 | DDC 616.07/54--dc23
LC record available at https://lccn.loc.gov/2016022758

Visit the Taylor & Francis Web site at
http://www.taylorandfrancis.com

and the CRC Press Web site at
http://www.crcpress.com

Contents

Series Preface

THE SERIES IN MEDICAL Physics and Biomedical Engineering describes the applications of physical sciences, engineering and mathematics in medicine and clinical research.

The series seeks (but is not restricted to) publications in the following topics:

- Artificial Organs

- Assistive Technology

- Bioinformatics

- Bioinstrumentation

- Biomaterials

- Biomechanics

- Biomedical Engineering

- Clinical Engineering

- Imaging

- Implants

- Medical Computing and Mathematics

- Medical/Surgical Devices

- Patient Monitoring

- Physiological Measurement

- Prosthetics

- Radiation Protection, Health Physics and Dosimetry

- Regulatory Issues

- Rehabilitation Engineering

- Sports Medicine

- Systems Physiology

- Telemedicine

- Tissue Engineering

- Treatment

The Series in Medical Physics and Biomedical Engineering is an international series that meets the need for up-to-date texts in this rapidly developing field. Books in the series range in level from introductory graduate textbooks and practical handbooks to more advanced expositions of current research.

The Series in Medical Physics and Biomedical Engineering is the official book series of the International Organization for Medical Physics.

THE INTERNATIONAL ORGANIZATION FOR MEDICAL PHYSICS

The International Organization for Medical Physics (IOMP) represents more than 18,000 medical physicists worldwide and has a membership of 80 national and 6 regional organizations, together with a number of corporate members. Individual medical physicists of all national member organisations are also automatically members.

The mission of IOMP is to advance medical physics practice worldwide by disseminating scientific and technical information, fostering the educational and professional development of medical physics and promoting the highest quality medical physics services for patients.

A World Congress on Medical Physics and Biomedical Engineering is held every three years in cooperation with the International Federation for Medical and Biological Engineering (IFMBE) and the International Union for Physics and Engineering Sciences in Medicine (IUPESM). A regionally based international conference, the International Congress of Medical Physics (ICMP) is held between world congresses. IOMP also sponsors international conferences, workshops and courses.

The IOMP has several programmes to assist medical physicists in developing countries. The joint IOMP Library Programme supports 75 active libraries in 43 developing countries, and the Used Equipment Programme coordinates equipment donations. The Travel Assistance Programme provides a limited number of grants to enable physicists to attend the world congresses.

IOMP co-sponsors the *Journal of Applied Clinical Medical Physics*. The IOMP publishes, twice a year, an electronic bulletin, *Medical Physics World*. IOMP also publishes e-Zine, an electronic news letter about six times a year. IOMP has an agreement with Taylor & Francis Group for the publication of the Medical Physics and Biomedical Engineering series of textbooks. IOMP members receive a discount.

IOMP collaborates with international organizations, such as the World Health Organisations (WHO), the International Atomic Energy Agency (IAEA) and other international professional bodies such as the International Radiation Protection Association (IRPA) and the International Commission on Radiological Protection (ICRP), to promote the development of medical physics and the safe use of radiation and medical devices.

Guidance on education, training and professional development of medical physicists is issued by IOMP, which is collaborating with other professional organizations in the development of a professional certification system for medical physicists that can be implemented on a global basis.

The IOMP website (www.iomp.org) contains information on all its activities, policy statements 1 and 2 and the 'IOMP: Review and Way Forward' which outlines all the activities of IOMP and plans for the future.

Preface

MODERN MEDICAL PRACTICE RELIES on a wide range of diagnostic imaging techniques capable of demonstrating in vivo anatomy, organ function and metabolic processes. Imaging technologies are now essential not only for their role in the diagnosis and staging of disease but also for guiding interventional and surgical procedures. Nuclear medicine has a long history in the use of radiation probe detectors for monitoring the organ uptake of administered tracers, starting with monitoring radio-iodine uptake in the thyroid gland. The diagnostic information obtained has been used to plan subsequent therapy using surgery, high-dose radiation or chemotherapy.

The first use of radiation detectors in surgical practice was recorded in 1949 when Selverstone and colleagues used Geiger–Muller tubes to detect the uptake of phosphorous-32 in 33 patients with brain tumours. Since then, new radiopharmaceuticals have been developed and adopted for clinical use and, together with the technological evolution of nucleonic instrumentation, have resulted in a sophisticated range of procedures for use at the bedside or within the operating room.

The use of gamma probe detectors has now become the standard practice during the assessment of patients with malignant diseases, such as in breast and head and neck cancer. Gamma probes are used for an increasing range of radioguided procedures with varying degrees of success. However, it is clear that the number of established bedside and surgical procedures is increasing steadily.

Despite the long-standing use of probe detectors, nuclear medicine and radionuclide radiology are mainly regarded as imaging modalities, and most major commercial equipment suppliers have concentrated on the production of large imaging systems such as PET and SPECT scanners for producing whole body and cross-sectional images. In recent years, these devices have been combined with x-ray CT scanners, resulting in hybrid systems with capabilities of fusing anatomical and functional images, with obvious advantages for the reporting physician or surgeon planning an operative approach. In comparison, the development of smaller gamma imaging systems for bedside and interventional use has been somewhat neglected, leaving university academic departments and small spin-out companies the task of equipment development and initial clinical evaluation. It is rather surprising that the development of miniature low-profile radiation detectors has enabled a new generation of gamma camera systems, but, at the time of writing, no major commercial equipment manufacturer has taken up this technology.

This idea for this book arose from a workshop held at the University of Leicester in February 2015 that was supported by the UK Science and Technology Facilities Council. The workshop was attended by a number of participants with interest and experience in the use of compact camera systems for intraoperative exploration. The following chapters cover the main developments that have been made over recent years, in the design and construction of compact and handheld gamma cameras, and provide in-depth information on a range of devices and their clinical and surgical use. For completeness, we have also included a chapter on radiopharmaceuticals and information on the more recent introduction of hybrid gamma–optical and fluorescent imaging, which we anticipate will make a substantial impact to this clinical field of endeavour.

We extend our gratitude to all the authors for their valuable contributions. We consider that this represents one of the most detailed and informative books available, covering the design and application of this new technology.

Alan Perkins
Nottingham, United Kingdom

John Lees
Leicester, United Kingdom

Acknowledgements

THE EDITORS THANK SARAH Bugby for her invaluable help in preparing the book and reviewing all the contributions. Special thanks to all our students, Layal Jambi, Mohammed Alqahtani, Aik Hao Ng, Numan Dawood and Bahadar Bhatia, and also to Bill McKnight, David Bassford, George Fraser and Elaine Blackshaw for their hard work, which has contributed to the success of the hybrid gamma camera described in Chapter 10. We also acknowledge the support from the UK Science and Technology Facilities Council who funded the workshop, Advances in Intraoperative Nuclear Imaging University of Leicester, 17 February 2015, for stimulating this work. We also thank Benn Bugby for proofreading the manuscript and for his help in the final preparation.

Editors

Professor Alan Perkins PhD, FIPEM, Hon FRCP is professor of medical physics in radiological sciences in the School of Medicine at the University of Nottingham and honorary consultant clinical scientist at Nottingham University Hospitals NHS Trust. He has more than 30 years of experience in medical physics, providing routine clinical service and research in nuclear medicine, radiopharmacology, drug delivery, ultrasound imaging and radiation protection. He is a former president of the British Nuclear Medicine Society, a former vice president of the Institute of Physics and Engineering in Medicine and an editor of the UK journal *Nuclear Medicine Communications.*

Professor John Lees PhD, CPhys, FInstP is a professor of imaging for life and medical sciences in the Department of Physics and Astronomy at the University of Leicester. He has more than 28 years of experience in radiation detectors and imaging systems. He was the UK detector scientist on the NASA/UK x-ray telescope Chandra, during which he pioneered the calibration of space instrumentation using synchrotron sources. Currently, he leads the BioImaging Unit in the Space Research Centre which encompasses fundamental detector research along with technology transfer from space instrumentation into medical and life sciences.

Contributors

J.R. Ballinger
Imaging Sciences
King's College
London, United Kingdom

E. Barranger
Service de Gynécologie
Hôpital Jean Verdier (Assistance
    Publique-Hôpitaux de Paris)
Bondy, France

A. Bricou
Service de Gynécologie
Hôpital Jean Verdier (Assistance
    Publique-Hôpitaux de Paris)
Bondy, France

S.L. Bugby
Space Research Centre
University of Leicester
Leicester, United Kingdom

Y. Charon
Laboratory Imagerie et
    Modélisation en Neurobiologie
    et Cancérologie
Centre National de la Recherche
    Scientifique
Orsay, France

K.-L. Cheung
School of Medicine
Royal Derby Hospital Centre
Nottingham University
Derby, United Kingdom

D.G. Darambara
Joint Department of Physics
Royal Marsden NHS Foundation
Trust
London, United Kingdom

L.J. de Wit-van der Veen
Antoni van Leeuwenhoek Hospital
The Netherlands Cancer Institute
Amsterdam, the Netherlands

M.-A. Duval
Laboratory Imagerie et
    Modélisation en Neurobiologie
    et Cancérologie
Paris, France

C. Hope
Health Education East Midlands
Royal Derby Hospital
Derby, United Kingdom

B. Janvier
Laboratory Imagerie et
    Modélisation en Neurobiologie
    et Cancérologie
Paris, France

K. Kerrou
Assistance Publique–Hôpitaux de
    Paris
Hôpital Tenon
Paris, France

F. Lefebvre
Laboratory Imagerie et
    Modélisation en Neurobiologie
    et Cancérologie
Centre National de la Recherche
    Scientifique
Orsay, France

M. McGurk
Department of Oral and
    Maxillofacial Surgery
Guy's Hospital
London, United Kingdom

L. Ménard
Laboratory Imagerie et
    Modélisation en Neurobiologie
    et Cancérologie
Paris, France

A.H. Ng
School of Medicine
University of Nottingham
Nottingham, United Kingdom

R. Parks
Nottingham University Hospital
    NHS Trust
Nottingham, United Kingdom

L. Pinot
Laboratory Imagerie et
    Modélisation en Neurobiologie
    et Cancérologie
Centre National de la Recherche
    Scientifique
Orsay, France

S. Pitre
Laboratory Imagerie et
    Modélisation en Neurobiologie
    et Cancérologie
Paris, France

B. Pouw
Antoni van Leeuwenhoek
    Hospital
The Netherlands Cancer Institute
Amsterdam, the Netherlands

C. Schilling
Department of Oral and
    Maxillofacial Surgery
Guy's Hospital
London, United Kingdom

M.P.M. Stokkel
Antoni van Leeuwenhoek
    Hospital
The Netherlands Cancer Institute
Amsterdam, the Netherlands

M.-A. Verdier
Laboratory Imagerie et
  Modélisation en Neurobiologie
  et Cancérologie
Centre National de la Recherche
  Scientifique
Orsay, France

T. Wendler
Faculty of Medicine
Department of Nuclear Medicine
Technische Universitaet Muenchen
Munich, Germany

CHAPTER 1

Small Field of View Gamma Cameras and Intraoperative Applications

Alan C. Perkins, A.H. Ng and John E. Lees

CONTENTS

1.1 DEVELOPMENT OF GAMMA CAMERA SYSTEMS

The medical specialty of nuclear medicine is based on the understanding of the physiological uptake and clearance of an administered radiolabelled tracer known as a radiopharmaceutical. Once administered to the patient, the radiopharmaceutical will concentrate in the organ or system of interest whilst emitting radiation due to the physical process of radioactive decay. For therapeutic purposes, radionuclides emitting beta or alpha particles are used to deposit energy at discrete sites in the body to achieve tumour sterilisation, whereas for imaging purposes, gamma-emitting radionuclides are used in conjunction with dedicated cameras and scanners that can image the spatial distribution of the radioactivity in the body.

Although early gamma scanners were around in the middle part of the last century, the concept of the gamma camera was invented by Hal Anger in 1952 whilst working at the Donner Laboratory in the University of California in Berkeley, USA. The early prototype consisted of a pinhole collimator and a sodium iodide flat intensifying screen placed in front of a photographic plate. The device demonstrated the ability to produce a clinically relevant image, by showing the gamma ray distribution emitted from the patient on the flat crystal. The sensitivity of the camera was low and a long exposure time and relatively high amounts of radioactivity were required to produce images. A few years later, Anger improved the prototype design by replacing the gamma ray detector with 4 in. diameter scintillator coupled to an array of seven photomultiplier tubes (PMTs). He also developed 'Anger arithmetic' to determine the position of the scintillation event that occurred in the crystal by comparing the photons collected in individual PMTs.

Anger's basic principle of the gamma camera has remained in use over the past 60 years. A gamma camera system consists of a collimator, scintillation detector, PMTs, preamplifiers, amplifier, pulse height analyser, X–Y positioning circuits and an image display with a recording or storage device. In modern systems, digital electronics and software are used to process the image data obtained from the PMT units. Basically, the incoming gamma rays from radiation sources will pass through the collimator and interact with the detector. The detector, usually made of a scintillation crystal, converts the energy from the gamma ray to a flash of light (scintillation event). The light photons are detected by an array of PMTs adjacent to the detector and the events are computed and mapped out as the gamma image display.

Over time gamma camera technology developed from the basic single-head detector camera to multi-head detectors mounted on a mechanical rotating gantry. The detector heads can scan over the patient to produce whole body image, and they can also rotate around the patient acquiring detected gamma events and producing cross-sectional images using a reconstruction software (in the same way as x-ray computed tomography [CT] image formation), a process known as single-photon emission CT (SPECT). These scanning techniques are employed in clinical nuclear medicine procedures such as brain and myocardial perfusion imaging. More recently, following the huge success of hybrid imaging modalities, positron emission tomography and CT (PET–CT), SPECT–CT camera systems have been introduced for the fusion of gamma and CT images to enable the overlay of physiological tracer uptake with the cross-sectional anatomical detail. These integrated modalities not only provide a method for localisation of the physiological data but also allow attenuation correction of gamma images by using the transmission data from the CT images.

1.2 SMALL FIELD OF VIEW GAMMA CAMERA

Most nuclear medicine procedures are normally carried out in the nuclear medicine department since gamma cameras are large bulky devices installed in purpose-built rooms. This requires the patient to attend the department in order to undergo the imaging investigation. As a consequence this restricts the types of patients who can access the services, for example, patients from the intensive care unit, operating theatre and accident and emergency department cannot always attend.

It is evident that mobile or portable gamma cameras may be of value in other hospital departments for bedside investigations. Amini et al. (2011) highlighted the usefulness of nuclear medicine techniques in the diagnosis of a range of conditions including chest pain, suspected pulmonary embolus, acute cholecystitis, gastrointestinal (GI) haemorrhage, acute scrotum and occult fractures. Hilton et al. (1994) reported that initial myocardial perfusion imaging was highly accurate to distinguish low- and high-risk conditions when carried out in the emergency department in patients presenting with typical angina and a normal or equivocal electrocardiogram.

The application of nuclear medicine techniques during surgical procedures has a long history. The concept of radioguided surgery (RGS) using a radiation gamma detector was first reported around 60 years ago by Selverstone et al. (1949). The early radiation detectors were bulky ionisation chambers, but the more recent surgical probe systems use sodium iodide or caesium iodide scintillation crystal devices for surgical and bedside investigations. In 1996, Perkins and Hardy described a range of surgical nuclear medicine procedures employing gamma probes (Perkins and Hardy 1996). These developments originated from the use of gamma probes for the detection of the uptake of ^{99m}Tc bone scanning agents for bone tumours, in particular osteoid osteoma, a benign condition that predominantly occurs in young children and adolescents. Probe systems were also used for a range of nuclear medicine investigations in the intensive care unit (Perkins and Hardy 1997). However, the more widespread introduction of gamma probes in surgical practice was subsequently driven by the routine introduction of sentinel lymph node biopsy (SLNB) procedures in the staging and management of breast cancer. This drove the evolution of gamma probe technology and the widespread acceptance of nuclear medicine procedures during surgery.

We are now at a stage in the evolution of intraoperative nuclear medicine procedures where high-resolution small field of view (SFOV) gamma cameras are becoming available for surgical use. For use in the operating theatre, gamma cameras must be mobile or portable. Ideally, they should be small enough to be handheld during surgery. As indicated by the name, SFOV gamma cameras are considerably smaller in comparison with conventional large field of view (LFOV) gamma cameras. Advances in detector technology have enabled the introduction of compact imaging systems with the cameras either being held on an articulated arm mounted on a small trolley or handheld by the operator during use.

1.3 MOBILE GAMMA CAMERAS

The typical LFOV gamma camera is designed for whole body imaging and SPECT. Whilst this enables a range of clinically important imaging investigations to be undertaken, it does limit the use of these cameras to patients who can be transferred to the nuclear medicine department. This does restrict access of patients who may be critically ill or have limited mobility. Smaller cameras with single-head detectors mounted on a mechanical trolley have been available over the past 20 years, and these have been used away from nuclear medicine departments to image patients in operating theatres and cardiac units. In 1973, Hurwitz et al.

reported preliminary experience with a 'portable' scintillation camera in the operating room or other locations previously believed to be inaccessible for diagnostic nuclear medicine (Hurwitz et al. 1973). The camera was a modified standard Anger scintillation camera with a ½ in. NaI(Tl) crystal 12 in. in diameter, mounted on a movable stand. In general, relatively large mobile camera systems have had limited clinical use; however, further technological developments in camera technology have enabled low-profile digital detectors for organ- and procedure-specific applications. The use of some of these systems is listed in Table 1.1.

Figure 1.1 shows a commercially available mobile gamma camera produced by Mediso Medical Imaging Systems, Budapest, Hungary. The camera detector comprises a thallium-doped sodium iodide NaI(Tl) scintillation crystal with a thickness of 6.5 mm backed by PMTs and is available in sizes of 230 mm × 210 mm, 260 mm × 246 mm and 300 mm × 300 mm. The system weighs 180 kg and the gantry is equipped with motorised height adjustment and interchangeable collimators. A separate computer console is used for image acquisition.

TABLE 1.1   Examples of Clinical Mobile Gamma Camera Systems

Camera	Detector	Nominal Field of View (mm)	Weight (kg)	Applications
Nucline™ TH (Mediso)	NaI(Tl), PMT	260 × 246	180	Cardiac, thyroid imaging (Mediso Medical Imaging Systems 2012)
ergo™ Imaging System (Digirad)	CsI(Tl), silicon photodiode	311 × 396	305	Cardiac, breast and other bedside imaging (Digirad Corporation 2014)
2020tc imager (Digirad)	CsI(Tl), silicon photodiode	212 × 212	386	Lung, thyroid, cardiac, mammoscintigraphy, bedside imaging (Digirad Corporation)
Cardiotom™ (Adolesco)	NaI(Tl), PMT (Sopha DS7)	400 diameter	320	Cardiac imaging (Adolesco 2001)
Picola–Scintron®	NaI(Tl), PMT	254 diameter	NA	Thyroid imaging (Medical Imaging Electronics)
Solo Mobile™ (DDD-Diagnostic A/S)	NaI, PMT	210 diameter	<270	Thyroid, parathyroid, cardiac, sentinel node imaging (DDD-Diagnostic A/S 2012)

Note: NaI(Tl), thallium-doped sodium iodide; CsI(Tl), thallium-doped caesium iodide.

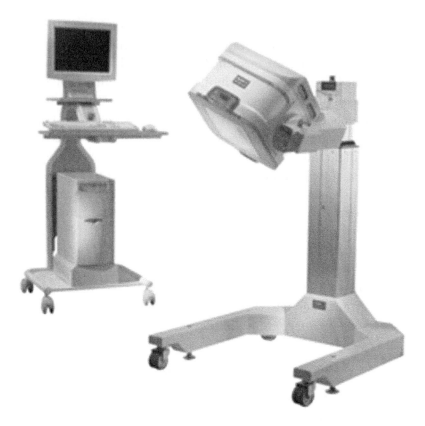

FIGURE 1.1 A mobile gamma camera – Mediso 'Nucline™ TH' – used for thyroid, heart, mammography and small organs imaging. (Taken from Mediso Medical Imaging Systems, 2012. Nucline™ TH. Retrieved 1 August 2014, http://www.mediso.com. With permission.)

The gamma camera shown in Figure 1.2 is a mobile digital gamma camera from Digirad Corporation, San Diego, CA. This has a field of view measuring 311 mm × 396 mm produced by a solid-state detector based on segmented thallium-doped caesium iodide (CsI(Tl)) with a silicon photodiode and is suitable for detecting gamma energies between 50 and 350 keV with the aid of the interchangeable collimators. The camera can be manoeuvred manually despite having a weight of 305 kg. Nuclear medicine procedures such as lung, liver, gastric emptying, GI bleed, renal, thyroid, brain flow, cardiac multiple-gated acquisition, molecular breast imaging, sentinel node, lymphoscintigraphy and veterinary studies may be carried out using the camera.

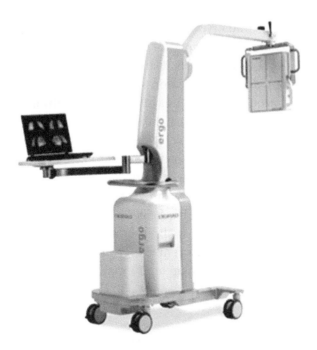

FIGURE 1.2 A solid-state mobile gamma camera – Digirad ergo™ Imaging System by Digirad. (Taken from Digirad Corporation, ergo™, 2014, Retrieved 1 August 2014, http://www.digirad.com. With permission.)

Another example of a mobile gamma camera is the Cardiotom™ system. This tomographic gamma camera system is based on a commercially available detector system attached with a rotatable parallel and slanted-hole collimator (ectomography). The camera weighs 320 kg and is primarily designed for cardiac studies at the patient bedside.

1.4 HANDHELD GAMMA CAMERAS

Advances in gamma camera detector design have led to the introduction of the handheld compact SFOV gamma camera. The concept of this handheld gamma imager was introduced and developed by Barber et al., in which the prototype devices are small enough to be inserted into the upper or lower GI tract to image the distribution of tumour-seeking radiotracers (Barber et al. 1984). There is now an increasing interest of the application of handheld SFOV gamma cameras in various clinical fields, but particularly for intraoperative imaging. Globally, there are a number of vibrant research initiatives underway exploring the development of robust, user-friendly and ergonomical SFOV gamma camera systems.

As with the LFOV cameras, the main features of the compact devices are the imaging detector, collimator, shielding, electronic readout circuit and image acquisition and image processing software. The design concepts provide flexibility and handheld use for additional areas of clinical investigation, and these cameras are not intended to replace the clinical applications undertaken by the standard LFOV gamma cameras. The size and weight of the camera head represent the most important characteristics of a SFOV camera in terms of ease of use and manoeuvrability. From the literature it has been found that the weight of SFOV camera heads used for clinical application was between 320 and 2490 g (Goertzen et al. 2013, Mathelin et al. 2007) whilst the detector size varied from 8.192 mm × 8.192 mm up to 127 mm × 127 mm (Aarsvod et al. 2006, Lees et al. 2014).

1.4.1 Camera Detectors

From reports published since 1997, most SFOV cameras have used either scintillation crystals (Fernández et al. 2004, Jung et al. 2009, Lees et al. 2011, Netter et al. 2009b, Olcott et al. 2004, Salvador et al. 2007, Soluri et al. 2005) or semiconductor (Abe et al. 2003, Crystal Photonics GmbH 2013, EuroMedical Instruments, Haneishi et al. 2010, Kopelman et al. 2005, Russo et al. 2011, Tsuchimochi et al. 2003) detectors. The scintillation crystals most commonly used are thallium- or sodium-doped caesium iodide (CsI(Tl) or CsI(Na)), thallium-doped sodium iodide (NaI(Tl)) and cerium-doped lutetium oxyorthosilicate (LSO). The semiconductor detector materials used in SFOV cameras are mainly cadmium telluride (CdTe) or cadmium zinc telluride (CdZnTe or CZT).

Patt et al. (1997) fabricated a 256-element mercuric iodide (HgI_2) detector array which was intended to be used as an intraoperative gamma camera. A novel SFOV gamma camera based on yttrium aluminium perovskite activated by cerium (YAP:Ce) coupled with intensified position-sensitive diode (IPSD) (Menard et al. 1998) has also been developed. The compact position-sensitive PMT (PSPMT) has been recognised as a key development in small camera design and these have been used in cameras with CsI(Tl) (Jung et al. 2009, Soluri et al. 2005), NaI(Tl) (Aarsvold et al. 2002, Olcott et al. 2004), cerium-doped orthosilicate of gadolinium (Gd_2SiO_5:Ce or GSO) (Salvador et al. 2007) and cerium-doped halide (LaBr3:Ce) scintillators (Netter et al. 2009b). Figure 1.3a gives an example of a 'small-sized

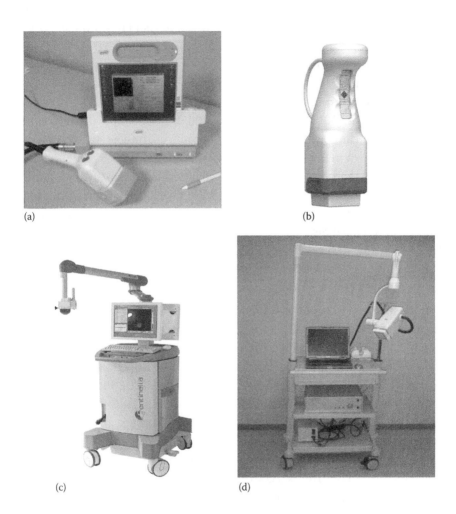

(a)

(b)

(c)

(d)

FIGURE 1.3   Examples of small field of view (SFOV) gamma cameras. (a) The IP Guardian2 portable γ-camera system with a scintillation detector. (Photo courtesy of Domenico Rubello, Rovigo, Italy.) (b) The CrystalCam imaging probe with semiconductor detector. (Courtesy of Thomas Barthel, Crystal Photonics GmbH, Berlin, Germany.) (c) Sentinella 102 SFOV gamma camera with pinhole collimator. (Courtesy of Victoria García, OncoVision, Spain.) (d) MGC 500 mini gamma camera with parallel hole collimator, mounted on an articulated arm and cart-based system. (Reprinted from *Phys. Med.*, 29(2), Tsuchimochi, M. and Hayama, K., Intraoperative gamma cameras for radioguided surgery: Technical characteristics, performance parameters, and clinical applications, 126–138. Copyright 2013, with permission from Elsevier.)

imaging probe', known as IP Guardian2 manufactured by Li-Tech. The detector of the SFOV gamma camera consists of CsI(Tl) scintillator coupled to PSPMT.

As for semiconductor-based cameras, several SFOV cameras have been produced using CdTe (EuroMedical Instruments, Jung et al. 2009, Russo et al. 2011, Tsuchimochi et al. 2003) or CZT (Abe et al. 2003, Crystal Photonics GmbH 2013, Kopelman et al. 2005). An example of an elegant and compact camera, the CrystalCam produced by Crystal Photonics GmbH, is shown in Figure 1.3b. This system is designed to be an 'imaging gamma probe' and has a range of tungsten collimators for use with different energy radionuclides. Lees et al. have developed a high-resolution SFOV gamma camera using a columnar CsI(Tl)-coated electron multiplying charge-coupled device (EMCCD) which is described in more detail in Chapters 9 and 10 (Lees et al. 2003, 2011).

1.4.2 Camera Collimators

The most common types of collimator used in SFOV camera are pinhole (Lees et al. 2011, Mettivier et al. 2003, Russo et al. 2011, Sanchez et al. 2004, 2006) and parallel hole (Abe et al. 2003, Crystal Photonics GmbH 2013, EuroMedical Instruments, Jung et al. 2009, Kopelman et al. 2005, Menard et al. 1999, Netter et al. 2009b, Olcott et al. 2007, Pitre et al. 2003, Salvador et al. 2007, Soluri et al. 2005, Tsuchimochi et al. 2003). These have been constructed from dense metal radiation absorbers, namely, lead or tungsten.

An example of a SFOV camera with pinhole collimator (Sentinella) is given in Figure 1.3c. This is a conventional camera system based on a pinhole collimator and a single CsI(Na) crystal backed by a PSPMT. A range of pinhole collimators is available for user selection. Figure 1.3d illustrates another SFOV gamma camera with parallel hole collimator, namely, mini gamma camera 500 (MGC500), produced by Acrorad Co. Ltd. The parallel hole collimator is a combination of 100 pieces of tungsten mesh sheets with a thickness of 0.1 mm, respectively. Each sheet has square openings measuring 1.2 mm × 1.2 mm with 0.2 mm pitch.

1.4.3 Camera Ergonomics and Design

As with any medical device, the construction and ergonomic design of compact cameras are important. Many of the prototype camera systems that have been produced are early constructs with crude metal casing and sharp edges; however, if these systems are to be used routinely, they will need to comply with medical device requirements satisfying the regulatory requirements, for example, CE marking in the EU and the 510(k) FDA requirement in the United States.

Although gamma cameras are not intended to be connected to patients, all devices must demonstrate appropriate electrical safety and electromagnetic compliance. For routine clinical use, the camera should be of solid construction and surfaces should be smooth and easily cleaned with sterile wipes. In most cases compact cameras would be placed in a sterile sheath or draped to ensure sterility during surgery and to enable repeated use with consecutive surgical cases. The camera head should be designed so that it can easily be placed in position over the patient and the user interface simple to operate for recording good quality scintigraphic images within a reasonably short acquisition time. Acquiring standard gamma images often takes a few minutes, although most surgeons would prefer to see images available within seconds rather than minutes.

It has been suggested that a FOV of approximately 50 mm × 50 mm is optimal for surgical application (Tsuchimochi and Hayama 2013). Table 1.2 shows the characteristics of the SFOV cameras that are currently available as research prototypes or in clinical use. The camera technical specifications are provided in Table 1.3.

1.5 CLINICAL APPLICATIONS AND PROCEDURES

The development of mobile gamma camera systems enables scintigraphic imaging whilst the patient is lying on a standard bed, in the ward, in private clinic, in intensive care facility or in the operating theatre. In particular, RGS procedures allow the surgeon to identify any uptake or 'suspicious' tissues in the body depending upon the specificity of the radiopharmaceutical administered. The early studies reported by Selverstone et al. at Harvard Medical School used Geiger–Müller tube devices to detect beta radiation emitter phosphorus-32 (^{32}P) in brain tumour patients (Selverstone et al. 1949). The counts in the area of suspected tumour and normal brain tissue were detected at various time intervals and various depths beneath the cerebral cortex to demarcate the tumour margins. In 1956, Harris et al., at the Oak Ridge Institute of Nuclear Studies Medical Hospital, reported the first use of a gamma detection probe system in RGS to detect the gamma radiation emitter

TABLE 1.2 Characteristics of Small Field of View Imaging Systems

Camera	Detector			Detector Head		Collimator		Energy Range (keV)	Clinical Application and Reference
	Type/ Material	Size (FOV) (xyz) (mm)	Matrix Size	Size (mm)	Weight (g)	Type	Material		
IOGC	Hgl2, electrodes	23.75 × 23.75 × 1	16 × 16	43 × 45	NA	Parallel hole	Tungsten	Tc-99m, Co-57, Am-241	NA (Patt et al. 1997)
HCGC	CsI(Tl), EMCCD	8.192 × 8.192 × 0.6	128 × 128	Ø 115 × 171 (L)	1500	Pinhole	Tungsten	30–140	Bone imaging, lacrimal drainage scan, thyroid imaging, lymphatic scan (Bugby et al. 2016, Lees et al. 2011, 2012)
POCI 2	CsI(Na), IPSD	Ø 40 × 3	50 × 50	Ø 95 × 90	1200	Parallel hole	Lead	Tc-99m, Co-57	SLN detection with breast cancer (Kerrou et al. 2011, Pitre et al. 2003)
TreCam	LaBr3:Ce, MAPMT	49 × 49 × 5	16 × 16	NA	2200	Parallel hole	Lead	Tc-99m, Co-57	SNOLL (Bricou et al. 2015, Netter et al. 2009a,b)
Mediprobe	CdTe:Cl	14.08 × 14.08 × 1	256 × 256	200 × 70 × 30	1500[a]	Pinhole	Tungsten	Tc-99m, Co-57	SLN detection with melanoma cancer (Russo et al. 2011)
eZ Scope	CZT	32 × 32 × 5	16 × 16	74 × 72 × 210	820	Parallel hole	NA	71–364	Intraoperative imaging for primary hyperparathyroidism (Abe et al. 2003, Fujii et al. 2011)
MGC500	CdTe	44.8 × 44.8	32 × 32	82 × 86 × 205	1400[a]	Parallel hole	Tungsten	550 keV max	Lymphoscintigraphy with oral cancer (Tsuchimochi et al. 2003, 2008, 2013)

(Continued)

TABLE 1.2 (Continued) Characteristics of Small Field of View Imaging Systems

| | Detector | | | Detector Head | | Collimator | | Energy Range (keV) | Clinical Application and Reference |
Camera	Type/ Material	Size (FOV) (mm)	Matrix Size	Size (mm)	Weight (g)	Type	Material		
IHGC	NaI(Tl), PSPMT	50 × 50 × 6	29 × 29	64 × 64 × 76	1100	Parallel hole	Lead	30–300	SLN detection with breast or melanoma cancer (Olcott et al. 2007, 2014)
IP	CsI(Tl), PSPMT	44.1 × 44.1 × 5	18 × 18	NA	1200	Parallel hole	Tungsten	50–250	SLN detection with breast cancer (Chondrogiannis et al. 2013, Ferretti et al. 2013, Trotta et al. 2007)
GE	CZT	40 × 40	16 × 16	150 (L)	1200	Parallel hole	NA	40–200	NA (Kopelman et al. 2005)
Sentinella 102	CsI(Na), PSPMT	40 × 40	300 × 300	80 × 90 × 1500	1000[a]	Pinhole	Lead	50–200	SLN detection with breast, penile and prostate cancer, intraoperative imaging for primary hyperparathyroidism (Brouwer et al. 2014, Fernández et al. 2004, General Equipment for Medical Imaging 2009, 2013, Scerrino et al. 2015, Vermeeren et al. 2010a, Vidal-Sicart et al. 2011)
GCOI	CsI(Tl), PSPMT	38.4 × 38.4 × 4	24 × 24	NA	NA	Parallel hole	NA	60–300	NA (Jung et al. 2007)

(Continued)

TABLE 1.2 (*Continued*) Characteristics of Small Field of View Imaging Systems

| Camera | Detector | | | Detector Head | | Collimator | | Energy Range (keV) | Clinical Application and Reference |
	Type/ Material	Size (FOV) (xyz) (mm)	Matrix Size	Size (mm)	Weight (g)	Type	Material		
CrystalCam	CZT	40 × 40	16 × 16	65 × 65 × 180	800	Parallel hole	Tungsten	30–250	SLN detection with breast and melanoma cancer (Crystal Photonics GmbH 2013, Engelen et al. 2015, Ozkan and Eroglu 2015, Zuhayra et al. 2015)
Minicam 2	CdTe	40 × 40	16 × 16	250 × 70 × 170	700	Parallel hole	Tantalum	30–200	SLN localisation (EuroMedical Intruments)
MRG15	CsI(Tl), SiPM	13.2 × 13.2 × 5	4 × 4	114 × 32 × 26	320	Parallel hole	NA	Tc-99m, Co-57	NA (Goertzen et al. 2013)
CarollRes	GSO:Ce, MAPMT	50 × 50	8 × 8	78 × 78 × 275	2490	Parallel hole	Lead	Tc-99m, Co-57	SLN detection with breast cancer (Mathelin et al. 2007, 2008)
Medica GammaCAM/ OR	NaI(Tl), PSPMT	127 × 127 × 6	56 × 56	NA	NA	Parallel hole	NA	Tc-99m	SLN detection with breast cancer (Aarsvold et al. 2006, Aarsvold et al. 2002)

Note: NA, information not available; L, length.
[a] Without collimator.

TABLE 1.3   Technical Specifications of Small Field of View Imaging Systems

Parameter	Intrinsic Spatial Resolution (FWHM, mm)	System Spatial Resolution (FWHM, mm)	System Sensitivity (cps/MBq)	Energy Resolution (%)
IOGC	5.3% ± 2.6%	5.2% ± 2.9%	NA	NA
POCI 2	2.3	3.2 at 0 mm	290 at 0 mm	32
HCGC	0.63	1.28	214	58
Mediprobe	NA	1.8 at 10 mm	6.6 at 50 mm	NA
eZ Scope	NA	2.2 at 0 mm	184.2 at 0 mm	8.6
SSGC	NA	1.59	300	6.9
IHGC	1.8	NA	270	12
IP	2.3	NA	211 at 50 mm	19.3
GE	5	NA	100	8
GCOI	NA	2.1	100	26
CrystalCam	NA	2.99 at 10 mm	922 at 10 mm	5.5
MRG15	3.46 at 1 mm	NA	149.7 at 50 mm	40.2
CarollRes	3 at 30 mm	10 at 30 mm	1000	45
Sentinella 102	1.8	5.4 at 30 mm	135 at 30 mm	15.9

Note: NA, information not available; FWHM, full width at half maximum.

from ^{131}I administered to a patient with a history of thyroid cancer (Harris et al. 1956). The gamma probe enabled the localisation of residual thyroid tissue prior to surgical resection.

In 2009, Povoski et al. reviewed the role of gamma probe technology and its use in the surgical management of various cancerous diseases, including but not limited to cutaneous malignancies, breast cancer, urologic malignancies and parathyroid disease (Povoski et al. 2009). The article also reported the impact of the technology in the surgical management of cancer patients including the provision of vital and real-time information to the surgeon on the localisation, demarcation and assessment of the surgical resection margins of the disease and the minimisation of the surgical invasiveness of many diagnostic and therapeutic procedures whilst maintaining optimum benefit to the patient.

With the development of the current technology, it is evident that SFOV handheld gamma cameras can offer potential clinical utility in any of the aforementioned areas of investigation previously carried out using gamma probes. There may be advantages or disadvantages in their use depending on the clinical or surgical application. For intraoperative use the size of the camera may limit some applications when compared to the use of the gamma probes which may provide greater ease of access

TABLE 1.4   Potential Clinical Applications of Small Field of View Gamma Cameras for Intraoperative Imaging

Specific Type(s) of Radioguided Surgery	Clinical Applications	References
Radioguided sentinel lymph node biopsy	Breast cancer	Chondrogiannis et al. (2013), Fernández et al. (2004), Kerrou et al. (2011), Mathelin et al. (2007, 2008), Menard et al. (1999), Olcott et al. (2014), Pitre et al. (2003), Salvador et al. (2007), Scopinaro et al. (2008), Soluri et al. (2006, 2007), Tsuchimochi et al. (2003), Vidal-Sicart et al. (2011, 2013), Wendler et al. (2010)
	Cutaneous malignancies	Fernández et al. (2004), Menard et al. (1999), Olcott et al. (2014), Riccardi et al. (2015), Russo et al. (2011), Sanchez et al. (2006)
	Head and neck malignancies	Tsuchimochi et al. (2008), Vermeeren et al. (2009a, 2010a)
	Urologic malignancies	Vermeeren et al. (2009b, 2010b, 2011)
	Gastrointestinal, gynaecologic, thoracic malignancies	
	Sarcoma	
Radioguided occult lesion localisation (ROLL) and sentinel node and occult lesion localisation (SNOLL)	Breast cancer	Bricou et al. (2015), Paredes et al. (2008)
Radioguided surgery (RGS)	Head and neck malignancies	Ferrer-Rebolleda et al. (2008), Fujii et al. (2011), Scerrino et al. (2015), Soluri et al. (2007)
	Neuroendocrine tumours	Soluri et al. (2007)
	Monitoring of isolated limb perfusion	Orero et al. (2009)
	Thoracic malignancies	
	Brain tumours	
	Bone lesions	
	Lymphoma	
Radioguided intraoperative margins evaluation (RIME)	Breast cancer	Vidal-Sicart et al. (2014)
Other	Brain death	Moron et al. (2009)

in small surgical incisions. Gamma probes give an audible signal which indicates the level of radiotracer uptakes at the site of uptake allowing the surgeon to home in on the target in the same way as surveying an area of ground with a metal detector. For most gamma-emitting radionuclides, the signal can be detected from sources at depth in the body and a camera dedicated for intraoperative imaging has the advantage of providing real-time static and dynamic images that may be recorded providing images for case notes, clinical audit and medicolegal purposes. Thus, it is evident that a camera may be used as a decisive tool during surgery (Povoski et al. 2009).

The range of clinical procedures include radioguided SLNB (RGSLNB), radioguided occult lesion localisation (ROLL), sentinel node and occult lesion localisation (SNOLL) and radioguided intraoperative margin evaluation (RIME). Table 1.4 provides a summary of the potential clinical applications of RGS that may be suitable for gamma probes and SFOV gamma cameras.

1.5.1 Radioguided Sentinel Lymph Node Biopsy

SLNB is a minimally invasive surgical technique which determines the likelihood of cancer spreading from a primary tumour and becoming metastatic throughout the lymphatic system. The standard protocol for the SLNB procedure uses a radiotracer (usually a small particulate, ^{99m}Tc-labelled nanocolloid) together with a blue dye to identify the location of the sentinel node. Numerous ^{99m}Tc-based agents have been utilised for radioguided SLN procedure for breast cancer. These include ^{99m}Tc-sulphur colloid, ^{99m}Tc-antimony sulphide, ^{99m}Tc-colloid human serum albumin (i.e., ^{99m}Tc-nanocolloid) and ^{99m}Tc-phytate. These are described in more detail in Chapter 6.

The amount of administered radioactivity typically varies between 3.7 and 370 MBq depending on study protocol and patient morphology (Giammarile et al. 2013, McCarter et al. 2001, Povoski et al. 2009, Van der Ent et al. 1999). Only a small fraction of the injected radiopharmaceutical transports and accumulates to SLN from the injection site. The mean percentage uptake of the injected dose per lymph node is approximately 0.9% (Viale et al. 1998) and 0.60% (Rink et al. 2001), whilst some studies have reported a range between 0.001% and 2.5% (Nieweg et al. 2001). In some centres, lymphoscintigraphy is performed using a standard gamma camera to image the distribution of the tracer before the surgery. Gamma imaging is commonly used prior to surgery and gamma probes are routinely used intraoperatively to identify the uptake of the radiotracer in the node.

Several research groups have described the use of SFOV gamma cameras in RGSLNB procedures (Fernández et al. 2004, Kerrou et al. 2011, Menard et al. 1999, Olcott et al. 2014, Pitre et al. 2003, Salvador et al. 2007, Scopinaro et al. 2008, Soluri et al. 2006, Tsuchimochi et al. 2003, Wendler et al. 2010). The most commonly reported application has been in the sentinel node localisation in patients with breast cancer (Fernández et al. 2004, Kerrou et al. 2011, Menard et al. 1999, Olcott et al. 2014, Pitre et al. 2003, Salvador et al. 2007, Scopinaro et al. 2008, Soluri et al. 2006, Wendler et al. 2010), followed by melanoma (Fernández et al. 2004, Menard et al. 1999, Olcott et al. 2014) and head and neck cancer (Vermeeren et al. 2010a). More recently, there has been grow- ing interest in the use of sentinel node biopsy procedures in patients with gynaecologic and urologic malignancies, for example, cervical and penile cancer.

The major advantages of using SFOV gamma cameras in RGSLNB are the high intraoperative detection rates in patients with difficult sentinel node localisation as assessed by preoperative lymphoscinti- gram (Vidal-Sicart et al. 2010). Also, it has been shown that cameras in combination with a gamma probe are useful to resolve SLNs that are deeply located or adjacent to other sites of high accumulation of radioactive tracers. In addition, SFOV gamma cameras may reduce sur- gery time by guiding the surgeon using real-time visualisation ensuring the complete clearance of the radioactive sentinel nodes or pathological uptake.

1.5.2 Radioguided Occult Lesion Localisation and Sentinel Node and Occult Lesion Localisation

There is an increasing incidence of the detection of non-palpable breast tumours due to advances in mammographic imaging techniques and the widespread use of the breast screening programmes. The accurate detec- tion of the non-palpable lesions pre- and intraoperatively as well as the precise localisation of the tumour are crucial for a high quality of surgical care. Previously, the most common method in guiding the surgical resec- tion of non-palpable breast cancer is wire-guided localisation where a guide wire is placed in situ under radiographic control and later is used to allow the surgeon to locate the tumour. ROLL is an increasingly popular technique, which was developed at the European Institute of Oncology, Milan, in 1998. The ROLL technique involves the injection of ^{99m}Tc to the breast tumour under the guidance of ultrasound or low-energy x-ray mammography. This can be combined with SLNB procedure to harvest SLN, which is also known as SNOLL. Radioactive seed localisation is another method to identify breast tumour with the insertion of a sealed radioactive ^{125}I seed source in the breast tissue to be excised. The gamma probe or camera is then used to locate the radioactive seed in order to guide the surgeon to the suspicious opacity or lesion during the surgical procedure.

1.5.3 Radioguided Surgery and Tumour Margin Evaluation

A range of surgical applications exist in addition to the more common procedures for RGS in breast cancer and sentinel node detection. Applications include the localisation of parathyroid adenomas and ectopic parathyroid tissue (Ortega et al. 2007) as well as the localisation of neuroendocrine tumours, bone lesions and isolated sites of infection (Vidal-Sicart et al. 2014). Other techniques involve the use of radiolabelled antibodies for the localisation of tumours, and there also is great interest in improving the assessment of tumour margins in order to fully remove tumour tissue whilst preserving the surrounding healthy tissues. The concept of guided intraoperative scintigraphic tumour targeting is described in the IAEA Human Health Series publication (IAEA 2014). From this report it is apparent that intraoperative nuclear imaging provides additional information, and the integration of these new techniques for 'virtual surgery' can minimise the complexity of surgical procedures.

1.6 CONCLUSION

The use of SFOV gamma cameras in surgery is expanding rapidly and is taking nuclear medicine procedures out of the traditional department setting. Ultimately, the utility of bedside imaging and RGS depends upon the specificity of the radiopharmaceuticals that are available. Further development of the radiotracers coupled with the technological advances in camera design will expand the range of procedures undertaken. The use of hybrid cameras combining optical and luminescent imaging with gamma imaging will further expand the range of surgical applications undertaken.

REFERENCES

Aarsvod, J.N., C.M. Greene, R.A. Mintzer et al. 2006. Intraoperative gamma imaging of axillary sentinel lymph nodes in breast cancer patients. *Physica Medica*, 21:76–79.

Aarsvold, J.N., R.A. Mintzer, C. Greene et al. November 2002. Gamma cameras for intraoperative localization of sentinel nodes: Technical requirements identified through operating room experience. In *IEEE Nuclear Science Symposium Conference Record*, Norfolk, Virginia. Vol. 2, pp. 1172–1176.

Abe, A., N. Takahashi, J. Lee et al. 2003. Performance evaluation of a hand-held, semiconductor (CdZnTe)-based gamma camera. *European Journal of Nuclear Medicine and Molecular Imaging*, 30(6):805–811.

Adolesco AB. 2001. 3D Mobile Gamma Camera – Cardiotom™. Retrieved 14 August 2015, from http://www.adolesco.se/.

Amini, B., C.B. Patel, M.R. Lewin, T. Kim and R.E. Fisher. 2011. Diagnostic nuclear medicine in the ED. *The American Journal of Emergency Medicine*, 29(1):91–101.

Barber, H.B., H.H. Barrett, W.J. Wild and J.M. Woolfenden. 1984. Development of small in-vivo imaging probes for tumor detection. *IEEE Transactions of Nuclear Science*, 31:599–604.

Bricou, A., M.A. Duval, L. Bardet et al. March 2015. Is there a role for a hand-held gamma camera (TReCam) in the SNOLL breast cancer procedure? *Quaterly Journal of Nuclear Medicine and Molecular Imaging*.

Brouwer, O.R., N.S. van den Berg, H.M. Mathéron et al. 2014. A hybrid radioactive and fluorescent tracer for sentinel node biopsy in penile carcinoma as a potential replacement for blue dye. *European Urology*, 65(3):600–609.

Bugby, S.L., J.E. Lees, A.H. Ng, M.S. Alqahtani and A.C. Perkins. 2016. Investigation of an SFOV hybrid gamma camera for thyroid imaging. *Physica Medica*, 32(1):290–296.

Chondrogiannis, S., A. Ferretti, E. Facci et al. 2013. Intraoperative hand-held imaging gamma-camera for sentinel node detection in patients with breast cancer feasibility evaluation and preliminary experience on 16 patients. *Clinical Nuclear Medicine*, 38(3) E132–E136.

Crystal Photonics GmbH. 2013. Handheld USB-gamma camera "CrystalCam". Retrieved 1 August 2014, from http://crystal-photonics.com/enu/products/cam-crystalcam-enu.htm.

DDD-Diagnostic A/S. 2012. SoloMobile. Retrieved 15 August 2015, from http://www.ddd-diagnostic.dk/PD.pdf.

Digirad Corporation. 2014. ergo™. Retrieved 1 August 2014, from http://www.digirad.com/cameras/ergo/.

Digirad Corporation. 2020tc Imager™. Retrieved August 14, 2015, from http://www.digirad.com/download/2020tc_s.pdf.

Engelen, T., B.M. Winkel, D.D. Rietbergen et al. 2015. The next evolution in radioguided surgery: Breast cancer related sentinel node localization using a freehandSPECT-mobile gamma camera combination. *American Journal of Nuclear Medicine and Molecular Imaging*, 5(3):233–245.

EuroMedical Instruments. Minicam 2. Retrieved 1 August 2014, from http://em-instruments.com.

Fernandez, M.M., J.M. Benlloch, J. Cerdá et al. 2004. A flat-panel-based mini gamma camera for lymph nodes studies. *Nuclear Instruments and Methods in Physics Research Section A: Accelerators, Spectrometers, Detectors and Associated Equipment*, 527(1):92–96.

Ferrer-Rebolleda, J., P. Sopena Novales, P. Estrems Navas et al. 2008. Contribution of a portable hand-held miniature gamma camera in surgical treatment of primary hyperparathyroidism. *Revista Española de Medicina Nuclear*, 27(2):124–127.

Ferretti, A., S. Chondrogiannis, A. Marcolongo and D. Rubello. 2013. Phantom study of a new hand-held gamma-imaging probe for radio-guided surgery. *Nuclear Medicine Communications*, 34(1):86–90.

Fujii, T., S. Yamaguchi, R. Yajima et al. 2011. Use of a handheld, semiconductor (cadmium zinc telluride)-based gamma camera in navigation surgery for primary hyperparathyroidism. *The American Surgery*, 77(6):690–693.

General Equipment for Medical Imaging S.A. 2009. Portable gamma camera: Sentinella 102. Retrieved December 17, 2015, from https://www.accessdata.fda.gov/cdrh_docs/pdf9/K092471.pdf.

General Equipment for Medical Imaging S.A. 2013. Sentinella 102. Retrieved October 16, 2013, from http://www.gem-imaging.com/descargas/produc-tos/sentinella102.pdf.

Giammarile, F., N. Alazraki, J.N. Aarsvold et al. 2013. The EANM and SNMMI practice guideline for lymphoscintigraphy and sentinel node localization in breast cancer. *European Journal of Nuclear Medicine and Molecular Imaging*, 40(12):1932–1947.

Goertzen, A.L., J.D. Thiessen, B. McIntosh, M.J. Simpson and J. Schellenberg. October 2013. Characterization of a handheld gamma camera for intraoper-ative use for sentinel lymph node biopsy. In *IEEE Nuclear Science Symposium and Medical Imaging Conference (NSS/MIC)*, Seoul, Korea. pp. 1–4.

Haneishi, H., H. Shimura and H. Hayashi. 2010. Image synthesis using a mini gamma camera and stereo optical cameras. *IEEE Transactions on Nuclear Science*, 57(3):1132–1138.

Harris, C.C., R.R. Bigelow, J.E. Francis, G.G. Kelley and P.R. Bell. 1956. A CsI (Tl)-crystal surgical scintillation probe. *Nucleonics*, 14:102–108.

Hilton, T.C., R.C. Thompson, H.J. Williams, R. Saylors, H. Fulmer and S.A. Stowers. 1994. Technetium-99m sestamibi myocardial perfusion imaging in the emergency room evaluation of chest pain. *Journal of the American College of Cardiology*, 23(5):1016–1022.

Hurwitz, S.R., W.L. Ashburn, J.P. Green and S.E. Halpern. 1973. Clinical appli-cations of a "portable" scintillation camera. *Journal of Nuclear Medicine*, 14(8):585–587.

International Atomic Energy Agency. 2014. Guided intraoperative scintigraphic tumour targeting (GOSTT): Implementing advanced hybrid molecular imaging and non-imaging probes for advanced cancer management. IAEA Human Health Series No. 29. Retrieved 5 February 2015, from http://www-pub.iaea.org/MTCD/Publications/PDF/Pub1648web-19833477.pdf.

Jung, J.H., Y. Choi, K.J. Hong et al. 2007. Development of a dual modality imaging system: A combined gamma camera and optical imager. In *IEEE Nuclear Science Symposium Conference Record*, Honolulu, Hawaii, pp. 3762–3765.

Jung, J.H., Y. Choi, K.J. Hong et al. 2009. Development of a dual modality imaging system: A combined gamma camera and optical imager. *Physics in Medicine and Biology*, 54(14):4547.

Kerrou, K., S. Pitre, C. Coutant et al. 2011. The usefulness of a preoperative compact imager, a hand-held γ-camera for breast cancer sentinel node biopsy: Final results of a prospective double-blind, clinical study. *Journal of Nuclear Medicine*, 52(9):1346–1353.

Kopelman, D., I. Blevis, G. Iosilevsky et al. 2005. A newly developed intraoperative gamma camera: Performance characteristics in a laboratory phantom study. *European Journal of Nuclear Medicine and Molecular Imaging*, 32(10):1217–1224.

Lees, J.E., D.J. Bassford, O.E. Blake, P.E. Blackshaw and A.C. Perkins. 2011. A high resolution Small Field Of View (SFOV) gamma camera: A columnar scintillator coated CCD imager for medical applications. *Journal of Instrumentation*, 6(12):C12033.

Lees, J.E., D.J. Bassford, O.E. Blake, P.E. Blackshaw and A.C. Perkins. 2012. A Hybrid Camera for simultaneous imaging of gamma and optical photons. *Journal of Instrumentation*, 7:P06009.

Lees, J.E., G.W. Fraser, A. Keay, D. Bassford, R. Ott and W. Ryder. 2003. The high resolution gamma imager (HRGI): A CCD based camera for medical imaging. *Nuclear Instruments and Methods in Physics Research Section A: Accelerators, Spectrometers, Detectors and Associated Equipment*, 513(1):23–26.

Lees, J.E., S.L. Bugby, B.S. Bhatia et al. 2014. A small field of view camera for hybrid gamma and optical imaging. *Journal of Instrumentation*, 9(12):C12020.

Mathelin, C., S. Salvador, D. Huss and J.L. Guyonnet. 2007. Precise localization of sentinel lymph nodes and estimation of their depth using a prototype intraoperative mini γ-camera in patients with breast cancer. *Journal of Nuclear Medicine*, 48(4):623–629.

Mathelin, C., S. Salvador, V. Bekaert et al. 2008. A new intraoperative gamma camera for the sentinel lymph node procedure in breast cancer. *Anticancer Research*, 28(5 B):2859–2864.

McCarter, M.D., H. Yeung, S. Yeh, J. Fey, P.I. Borgen and H.S. Cody III. 2001. Localization of the sentinel node in breast cancer: Identical results with same-day and day-before isotope injection. *Annals of Surgical Oncology*, 8(8):682–686.

Medical Imaging Electronics. Picola Scintron®. Retrieved 15 August 2015, from http://www.mieamerica.com/.

Mediso Medical Imaging Systems. 2012. Nucline™ TH. Retrieved 1 August 2014, from http://www.mediso.com/.

Menard, L., Y. Charon, M. Solal et al. 1998. POCI: A compact high resolution γ camera for intra-operative surgical use. *IEEE Transactions on Nuclear Science*, 45(3):1293–1297.

Menard, L., Y. Charon, M. Solal et al. 1999. Performance characterization and first clinical evaluation of a intra-operative compact gamma imager. *IEEE Transactions on Nuclear Science*, 46(6):2068–2074.

Mettivier, G., M.C. Montesi and P. Russo. 2003. Design of a compact gamma camera with semiconductor hybrid pixel detectors: Imaging tests with a pinhole collimator. *Nuclear Instruments and Methods in Physics Research Section A: Accelerators, Spectrometers, Detectors and Associated Equipment*, 509(1):321–327.

Moron, C.C., P.A.D. Perez, T.C. Molina et al. 2009. Brain perfusion image with a portable mini-gamma camera (Sentinella (R)) in brain death. *Revista Espanola De Medicina Nuclear*, 28(2):83–84.

Netter, E., L. Pinot, L. Menard et al. October 2009a. Designing the scintillation module of a pixelated mini gamma camera: The spatial spreading behaviour of light. In *IEEE Nuclear Science Symposium Conference Record (NSS/MIC)*, Orlando, FL, pp. 3300–3302.

Netter, E., L. Pinot, L. Menard et al. October 2009b. The Tumor Resection Camera (TReCam), a multipixel imaging probe for radio-guided surgery. In *IEEE Nuclear Science Symposium Conference Record (NSS/MIC)*, Orlando, Florida. pp. 2573–2576.

Nieweg, O.E., P.J. Tanis, J.D.H. de Vries, R.A. Valdes Olmos, B.B.R. Kroon and C.A. Hoefnagel. June 2001. The sentinel node in melanoma: Present controversies. In *The 48th Society of Nuclear Medicine Annual Meeting*, Toronto, Canada. pp. 94–103.

Olcott, P.D., F. Habte, C.S. Levin and A.M. Foudray. 2004. Characterization of performance of a miniature, high sensitivity gamma-ray camera. *IEEE Nuclear Science Symposium and Medical Imaging Conference Record*, Rome, Italy, pp. 3997–4000.

Olcott, P., G. Pratx, D. Johnson, E. Mittra, R. Niederkohr and C.S. Levin. 2014. Clinical evaluation of a novel intraoperative handheld gamma camera for sentinel lymph node biopsy. *Physica Medica*, 30(3):340–345.

Olcott, P.D., F. Habte, A.M. Foudray and C.S. Levin. 2007. Performance characterization of a miniature, high sensitivity gamma ray camera. *IEEE Transactions on Nuclear Science*, 54(5):1492–1497.

Orero, A., S. Vidal-Sicart, N. Roe et al. 2009. Monitoring system for isolated limb perfusion based on a portable gamma camera. *Nuklearmedizin-Nuclear Medicine*, 48(4):166–172.

Ortega, J., J. Ferrer-Rebolleda, N. Cassinello and S. Lledo. 2007. Potential role of a new hand-held miniature gamma camera in performing minimally invasive parathyroidectomy. *European Journal of Nuclear Medicine and Molecular Imaging*, 34(2):165–169.

Ozkan, E. and A. Eroglu. 2015. The utility of intraoperative handheld gamma camera for detection of sentinel lymph nodes in melanoma. *Nuclear Medicine and Molecular Imaging*, 49(4):318–320.

Paredes, P., S. Vidal-Sicart, G. Zanon et al. 2008. Radioguided occult lesion localisation in breast cancer using an intraoperative portable gamma camera: First results. *European Journal of Nuclear Medicine and Molecular Imaging*, 35(2):230–235.

Patt, B.E., M.P. Tornai, J.S. Iwanczyk, C.S. Levin and E.J. Hoffman. 1997. Development of an intraoperative gamma camera based on a 256-pixel mercuric iodide detector array. *IEEE Transactions on Nuclear Science*, 44(3):1242–1248.

Perkins, A.C. and J.G. Hardy. 1996. Intra-operative nuclear medicine in surgical practice. *Nuclear Medicine Communications*, 17(12):1006–1015.

Perkins AC, Yeoman P, Hindle AJ et al. 1997. Bedside nuclear medicine investigations in the intensive care unit. *Nuclear Medicine Communications*, 18(3):262–268.

Pitre, S., L. Ménard, M. Ricard, M. Solal, J.R. Garbay and Y. Charon. 2003. A hand-held imaging probe for radio-guided surgery: Physical performance and preliminary clinical experience. *European Journal of Nuclear Medicine and Molecular Imaging*, 30(3):339–343.

Povoski, S.P., R.L. Neff, C.M. Mojzisik et al. 2009. A comprehensive overview of radioguided surgery using gamma detection probe technology. *World Journal of Surgical Oncology*, 7(1):11.

Riccardi, L., M. Gabusi, M. Bignotto et al. 2015. Assessing good operating conditions for intraoperative imaging of melanoma sentinel nodes by a portable gamma camera. *Physica Medica*, 31(1):92–97.

Rink, T., T. Heuser, H. Fitz, H.J. Schroth, E. Weller and H.H. Zippel. 2001. Lymphoscintigraphic sentinel node imaging and gamma probe detection in breast cancer with Tc-99m nanocolloidal albumin: Results of an optimized protocol. *Clinical Nuclear Medicine*, 26(4):293–298.

Russo, P., A.S. Curion, G. Mettivier et al. 2011. Evaluation of a CdTe semiconductor based compact gamma camera for sentinel lymph node imaging. *Medical Physics*, 38(3):1547–1560.

Salvador, S., V. Bekaert, C. Mathelin, J.L. Guyonnet and D. Huss. 2007. An operative gamma camera for sentinel lymph node procedure in case of breast cancer. *Journal of Instrumentation*, 2(07):P07003.

Sanchez, F., J.M. Benlloch, B. Escat et al. 2004. Design and tests of a portable mini gamma camera. *Medical Physics*, 31(6):1384–1397.

Sánchez, F., M.M. Fernández, M. Giménez et al. 2006. Performance tests of two portable mini gamma cameras for medical applications. *Medical Physics*, 33(11):4210–4220.

Scerrino, G., S. Castorina, G.I. Melfa et al. 2015. The intraoperative use of the mini-gamma camera (MGC) in the surgical treatment of primary hyperparathyroidism Technical reports and immediate results from the initial experience. *Annali italiani di chirurgia*, 86:212–218.

Scopinaro, F., A. Tofani, G. di Santo et al. 2008. High-resolution, hand-held camera for sentinel-node detection. *Cancer Biotherapy & Radiopharmaceuticals*, 23(1):43–52.

Selverstone, B., W.H. Sweet and C.V. Robinson. 1949. The clinical use of radioactive phosphorus in the surgery of brain tumors. *Annals of Surgery*, 130(4):643.

Soluri, A., R. Massari, C. Trotta et al. 2005. New imaging probe with crystals integrated in the collimator's square holes. *Nuclear Instruments and Methods in Physics Research Section A: Accelerators, Spectrometers, Detectors and Associated Equipment*, 554(1):331–339.

Soluri, A., R. Massari, C. Trotta et al. 2006. Small field of view, high-resolution, portable γ-camera for axillary sentinel node detection. *Nuclear Instruments and Methods in Physics Research Section A: Accelerators, Spectrometers, Detectors and Associated Equipment*, 569(2):273–276.

Soluri, A., C. Trotta, F. Scopinaro et al. 2007. Radioisotope guided surgery with imaging probe, a hand-held high-resolution gamma camera. *Nuclear Instruments & Methods in Physics Research Section A-Accelerators Spectrometers Detectors and Associated Equipment*, 583(2–3):366–371.

Trotta, C., R. Massari, N. Palermo et al. 2007. New high spatial resolution portable camera in medical imaging. *Nuclear Instruments & Methods in Physics Research Section A-Accelerators Spectrometers Detectors and Associated Equipment*, 577(3):604–610.

Tsuchimochi, M., H. Sakahara, K. Hayama et al. 2003. A prototype small CdTe gamma camera for radioguided surgery and other imaging applications. *European Journal of Nuclear Medicine and Molecular Imaging*, 30(12):1605–1614.

Tsuchimochi, M., K. Hayama, T. Oda et al. 2008. Evaluation of the efficacy of a small CdTe gamma-camera for sentinel lymph node biopsy. *Journal of Nuclear Medicine*, 49(6):956–962.

Tsuchimochi, M. and K. Hayama. 2013. Intraoperative gamma cameras for radioguided surgery: Technical characteristics, performance parameters, and clinical applications. *Physica Medica*, 29(2):126–138.

Van der Ent, F.W.C., R.A.M. Kengen, H.A.G. Van der Pol and A.G.M. Hoofwijk. 1999. Sentinel node biopsy in 70 unselected patients with breast cancer: Increased feasibility by using 10 mCi radiocolloid in combination with a blue dye tracer. *European Journal of Surgical Oncology*, 25(1):24–29.

Vermeeren, L., W.M. Klop, M.W. van den Brekel et al. 2009a. Sentinel node detection in head and neck malignancies: Innovations in radioguided surgery. *Journal of Oncology*, 2009:681746.

Vermeeren, L., R.A.V. Olmos, W. Meinhardt et al. 2009b. Intraoperative radioguidance with a portable gamma camera: A novel technique for laparoscopic sentinel node localisation in urological malignancies. *European Journal of Nuclear Medicine and Molecular Imaging*, 36(7):1029–1036.

Vermeeren, L., R.A.V. Olmos, W. Meinhardt and S. Horenblas. 2011. Intraoperative imaging for sentinel node identification prostate carcinoma: Its use in combination with other techniques. *Journal of Nuclear Medicine*, 52(5):741–744.

Vermeeren, L., R.A. Valdes Olmos, W.M.C. Klop, A.J. Balm and M.W. van den Brekel. 2010a. A portable γ-Camera for intraoperative detection of sentinel nodes in the head and neck region. *Journal of Nuclear Medicine*, 51(5):700–703.

Vermeeren, L., S.H. Muller, W. Meinhardt and R.A.V. Olmos. 2010b. Optimizing the colloid particle concentration for improved preoperative and intraoperative image-guided detection of sentinel nodes in prostate cancer. *European Journal of Nuclear Medicine and Molecular Imaging*, 37(7):1328–1334.

Viale, P.C.G., U. Veronesi and G. Paganelli. 1998. Lymphoscintigraphy and radioguided biopsy of the sentinel axillary node in breast cancer. *Journal of Nuclear Medicine*, 39:2080–2084.

Vidal-Sicart, S., P. Paredes, G. Zanón et al. 2010. Added value of intraoperative real-time imaging in searches for difficult-to-locate sentinel nodes. *Journal of Nuclear Medicine*, 51(8):1219–1225.

Vidal-Sicart, S., L. Vermeeren, O.Sola et al. 2011. The use of a portable gamma camera for preoperative lymphatic mapping: A comparison with a conventional gamma camera. *European Journal of Nuclear Medicine and Molecular Imaging*, 38(4):636–641.

Vidal-Sicart, S., F. Giammarile, G. Mariani and R.A. Valdés Olmos. 2013. Pre- and intra-operative imaging techniques for sentinel node localization in breast cancer. *Imaging in Medicine*, 5(3):275–291.

Vidal-Sicart, S., M.E. Rioja, P. Paredes, M.R. Keshtgar and R.A. Valdés Olmos. 2014. Contribution of perioperative imaging to radioguided surgery. *Quarterly Journal of Nuclear Medicine and Molecular Imaging*, 58(2):140–160.

Wendler, T., K. Herrmann, A. Schnelzer et al. 2010. First demonstration of 3-D lymphatic mapping in breast cancer using freehand SPECT. *European Journal of Nuclear Medicine and Molecular Imaging*, 37(8):1452–1461.

Zuhayra, M., S. Dierck, M. Marx et al. 2015. Is the portable gamma camera "Crystal Cam" equal to the conventional gamma camera for the detection of SLNs of malignant melanoma. *Journal of Nuclear Medicine*, 56(supplement 3):1866.

CHAPTER 2

Detector Design for Small Field of View (SFOV) Nuclear Cameras

D.G. Darambara

CONTENTS

2.1 INTRODUCTION

Cancer patient management depends on the accurate localisation and staging of a tumour as well as on its removal. Imaging has traditionally played a key role in the early diagnosis, accurate staging and treatment optimisation of various cancers. It is essential to develop and implement techniques capable of detecting and identifying small and deep tumours from the adjacent background with significantly high specificity, while similarly achieving high detection sensitivity, but also facilitating their effective surgical removal. There is a pressing need therefore for compact, small field of view (SFOV), task-specific nuclear cameras with high spatial resolution being of prime importance. Such a camera will be able overall to perform better than a standard gamma camera provided that it is placed in close proximity to the target volume/organ to be imaged and the background uptake of the radiotracer is fairly uniform or the camera itself has the ability to correctly quantify and deduct the adjacent background.

There is often the necessity to image small organs, for example, thyroid or lymph nodes. However, it is very difficult to position the general-purpose gamma cameras (Anger 1958, Madsen 2007, Petersen and Furenlid 2011) very close to the organ of interest, as required, due to their bulky and heavy design. The conventional nuclear cameras receive a significant fraction of background activity from other organs of the body, and in addition only certain views can be obtained, which may not include the most suitable one. Furthermore, the standard clinical gamma cameras suffer from inadequate spatial and energy resolutions, relatively low sensitivity to the disease and to the detection of small diameter tumours and degradation of contrast due to scatter and adjacent background activity. Finally, the cost per study is relatively high.

Consequently, the radionuclide-based detection and localisation of small tumours by employing conventional gamma cameras have several limitations: low tumour uptake of radiotracers, low overall detection sensitivity, possible contamination of significant fraction of counts originating from the targeted tumour tissue by counts coming from adjacent background activity in neighbouring tissues and degradation of contrast and spatial resolution due to Compton scatter.

Such problems limit the usefulness of general-purpose nuclear cameras for imaging small organs but can be remedied, and hence, the radionuclide detection can be enhanced by the use of counting (non-imaging) and particularly imaging intraoperative probes and/or dedicated, high-resolution (of less than 1 mm) SFOV gamma cameras capable of detecting gamma rays over a wide range of clinically useful energies (radionuclides) with the highest possible efficiency and localising them as accurately as possible (Halkar and Aarsvold 1999, Heller and Zanzonico 2011, Hoffman et al. 1999, 2004, Levin 2004, Zanzonico and Heller 2000). A SFOV nuclear camera can image a target volume, or a specific tissue region, or an organ of the body in close proximity achieving significantly high intrinsic spatial resolution and at various views achieving improved detection sensitivity by avoiding unwanted background events from other body parts. There are several advantages to using SFOV gamma cameras which demonstrate their potential importance: compact design, lightweight, high portability, flexibility, close-proximity imaging, increased sensitivity and spatial resolution, applicability in low-contrast environment and high image quality at reasonably low cost.

In recent years, due to advances in detector technologies, the slowly developing field of SFOV gamma cameras with high position resolution has been the centre of attention, and increased interest has led to a few of them already having reached the market and been made commercially available (Tsuchimochi and Hayama 2013). These cameras offer direct visualisation of the spatial distribution of the detected radioactivity and can find various applications as medical devices assisting in the diagnosis and treatment of cancer. They provide real-time information regarding location and extent of malignancy and facilitating imaging to be carried out in the operating theatre and at the bedside, such as intraoperative radioguided surgery, tumour localisation and resection, imaging of targeted small organs/tissues (e.g. thyroid, parathyroid, etc.), accurate sentinel lymph node identification and localisation in breast cancer and other types of cancer surgeries and scintimammography (Goertzen et al. 2013, Keshtgar and Ell 1999, Kopelman et al. 2005, Mueller et al. 2003, Olcott et al. 2014, Povoski et al. 2009, Raylman and Wahl 1994, Sarikaya et al. 2008, Sauret et al. 1999, Schneebaum et al. 1999, Strong et al. 2008, Valdes et al. 2010).

Various detector configurations for high-resolution SFOV gamma cameras are described in this chapter and their advantages and disadvantages are discussed. Important factors in the design, geometry and performance of SFOV gamma cameras with high-resolution capabilities are also identified and reviewed.

2.2 DETECTOR DESIGN CONFIGURATIONS FOR SFOV NUCLEAR CAMERAS

Most designs of the SFOV gamma cameras are application specific. As such, the overall dimensions of these cameras, and particularly the geometric components of the detector and the thickness of the shielding, depend on the application for which the camera will be employed. The SFOV nuclear cameras come in various shapes and numerous designs.

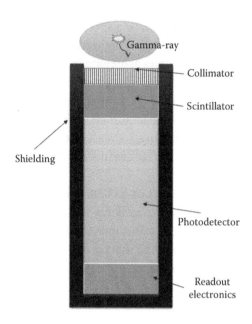

FIGURE 2.1   Schematic cross section of a small field of view nuclear camera.

A SFOV gamma camera consists of a detector, a collimator and readout electronics including bias, energy and position signal processing, pulse height analysis and gating and background shielding integrated in a single compact device (Figure 2.1).

A typical collimator (Hartsough et al. 1995) and any shielding on the back and sides of a camera are appropriately designed to reduce the number of gamma rays that strike the detector and, therefore, to protect the detector from any background radiation. The collimation is an important factor as it affects spatial resolution and geometrical sensitivity, while the shielding reduces scattering and undesired activity coming from outside the effective FOV, hence, decreasing image degradation. The effective FOV, which is the target volume of tissue from which the gamma rays depart and enter the detector, has the shape of a 3D cone diverging outward from the collimator and increases as the distance from the collimator increases. A long collimator with a narrow aperture decreases the FOV and, hence, decreases the sensitivity, while increasing the spatial resolution. Parallel and pinhole designs are the two most common collimators currently employed in SFOV gamma cameras. These collimators may be implemented in an interchangeable format and are made of high-Z materials, such as lead, tungsten, gold and platinum.

The detector configuration mainly implemented in SFOV gamma cameras consists of a combination of a scintillator crystal (single continuous or segmented) optically coupled to a photomultiplier tube (PMT). More recently, significant advances in photomultipliers and solid-state detector technologies led to the development of position-sensitive PMTs (PSPMTs) (Kim et al. 2000, Matthews 1997, Sanchez et al. 2004, Williams et al. 2000), silicon photodiodes (Levin et al. 1996) and drift detectors (Fiorini et al. 1999), charge coupled devices (Bugby et al. 2014, Lees et al. 2011), semiconductor photodetectors (Lewellen 2008, 2010, Renker and Lorenz 2009) and pixelated compound semiconductors (Darambara and Todd-Pokropek 2002, Knoll et al. 2014, Raylman 2001) for medical imaging applications (Levin 2003), which have also been exploited as SFOV gamma cameras. The earliest SFOV nuclear cameras were based on conventional scintillator crystals, such as NaI(Tl), CsI(Na) and CsI(Tl), coupled to a PMT with optical fibres covering an FOV of 1.5–2.5 cm in diameter. Later, SFOV gamma cameras were based on a mosaic (2D array) of scintillator crystals coupled to a PSPMT with a larger effective FOV of 2–10 cm in diameter. More recent SFOV nuclear cameras consist of semiconductor detectors such as cadmium telluride (CdTe) (Tsuchimochi et al. 2003, 2008), cadmium zinc telluride (CdZnTe or CZT) (Abe et al. 2003, Mueller et al. 2003, Parnham et al. 2001) and mercuric iodide (HgI_2) (Barber et al. 1991, Kwo et al. 1991, Patt et al. 1997) with an FOV of 1.5 cm × 1.5 cm^2 to 5 cm × 5 cm^2. The very first SFOV nuclear cameras were handheld devices with a very small and limited FOV, and thus, it was required that a large number of images be acquired to cover the anatomical area of interest making it difficult to hold the camera still for an extended period.

The overall performance of a SFOV gamma camera depends on the technical properties of the camera, that is, its detector overall sensitivity (absolute and geometrical), spatial resolution, energy resolution, collimation and scatter rejection capability, but also on its size and weight and finally on the nuclear characteristics (emitted photon energy and half-life) of the administered radiopharmaceuticals.

2.3 KEY PERFORMANCE PARAMETERS FOR SFOV NUCLEAR CAMERAS

Radiation imaging detectors are quantitatively characterised and optimised by various operational parameters that impact their performance (Bhatia et al. 2015, Cherry et al. 2003, Guerra et al. 2009, Knoll 2000, Levin 2004, Myronakis and Darambara 2011, Olcott et al. 2007, Zanzonico and

Heller 2000). There are several fundamental factors that have been proven essential and useful in characterising, optimising and comparing the overall imaging performance of SFOV gamma cameras.

2.3.1 Detection Efficiency

Detection efficiency as a measure of a detector's ability to detect useful signals generated by radiation incident on it, and is very important for quantitative measurements and depends on the source–detector geometry and the detector properties. The absolute efficiency of a detector is defined as the ratio of the number of photons detected to the number of photons emitted by the source. Detector sensitivity is the ratio between count rates and activity of a source at a specific distance and decreases with distance. The collimator thickness may be reduced or a wide energy window may be implemented to improve and increase the detection sensitivity. As a result though, the scattered counts will also be increased leading to degradation of spatial resolution and contrast and, thus, degradation of the camera's ability to localise a tumour. The collimator geometrical specifications should be thoroughly designed to achieve high detection efficiency. Further, to achieve statistically significant counts in a short period of measurement time and in a relatively large area of interest, a SFOV camera should exhibit high geometric and intrinsic detection efficiencies. The geometric efficiency is the ratio of the number of photons that strike the detector to the number of photons emitted from the source in all directions and depends on the detector size and the detector–source geometry. The intrinsic efficiency is the ratio of photons recorded to the number of photons incident on the detector and depends on the detector material, photon energy and detector thickness. It is almost independent of the geometry except at very short distances between detector and source. For a semiconductor-based SFOV gamma camera over the energy range of interest, the detector should be adequately thick, and the semiconductor material should have high atomic number Z and high density to achieve high stopping power and, thus, high detection efficiency. The higher the detection efficiency of a camera, the better. High sensitivity is important for detecting lesions with a diameter at least equal to the size of the detector of a SFOV camera.

2.3.2 Energy Resolution

Energy resolution quantifies the ability of a gamma-ray detector to identify and differentiate photons of different energy. It is a measure of the detector's ability to resolve two closely located peaks in energy. The photopeak of an energy spectrum represents the total number of pulses generated by the interactions in the detector and approximates a Gaussian-shaped curve. The energy resolution can be portrayed as the width of the photopeak of an energy spectrum. The peak width is determined by the statistical fluctuations associated with the charge production in the detector and contributions from the pulse-processing electronics. The energy resolution is conventionally expressed as a percentage of full width at half maximum (FWHM) of a gamma-ray energy spectrum with specific photopeak energy as $FWHM(\%) = (\Delta E/E_\gamma) \times 100\%$, where ΔE is the width and E_γ the energy of the photopeak. High energy resolution allows the use of narrow energy windows for better discriminations and reduction of energy-dependent background effects as well as more effective scatter rejection, which can degrade the final image quality. A good energy resolution is also important in a case where more than one radionuclides of different energies are imaged. Good energy linearity over a wide energy range of interest and good uniformity of the detector's energy response without significant variations across the entire effective FOV are essential for eliminating any artefacts that will degrade the imaging performance of the SFOV camera. Narrower photopeak widths imply high energy resolution. The smaller the FWHM energy resolution, the better. Semiconductor detectors have <1% energy resolution, while scintillators are 5%–10%, for example, the energy resolution of a CdZnTe detector is 1%–2% at 662 keV, while for the NaI scintillator is 7%.

2.3.3 Spatial Resolution

Spatial resolution is the ability of a gamma-ray detector to accurately determine the position of a source and to separate two sources which are located near to each other, and also to resolve finer structures. High spatial resolution of the detector helps to detect, count and localise radiation and to resolve uptake from the adjacent background. The system spatial resolution of a SFOV gamma camera is determined by the collimator geometrical properties and the detector material and pixel size and/or segmentation (fine intrinsic spatial resolution). Good spatial linearity and good spatial response uniformity without variations in the detector response within the small FOV are significant for minimising non-uniformities, artefacts and distortions of important structures, which degrade the imaging performance of a SFOV gamma camera.

High spatial resolution is essential for the detection of very small and/ or deep tumours.

There is a trade-off between spatial resolution and sensitivity, which are inversely related. High sensitivity is required for detecting a small amount of activity coming from the target tissue volume. For identifying a small amount of activity adjacent to a larger non-specific high activity, excellent spatial resolution is necessary. Sensitivity, energy resolution and spatial resolution also affect the *contrast* of a SFOV gamma camera, which differentiates high count rates in a target area from the low count rates in a neighbouring background area. For sentinel node detection, it is important to have excellent spatial resolution, while sensitivity and energy resolution are less crucial. For radioguided tumour surgery, it is primarily important to have high sensitivity, but also good energy resolution and fine spatial resolution within high–count rate areas are equally significant. SFOV gamma cameras with high energy resolution allow the implementation of narrow energy windows for more efficient rejection of Compton scatter, while at the same time allow adequately good sensitivity to be maintained.

Any signal recorded by a detector has a finite processing time, which determines the count rate capability of the detector. A detector has a characteristic dead time, which is the finite amount of time required to record and process an event during which no other event can be recorded. Any events that occur during this time are lost. Any detector has limits to the rate at which events may be processed and recorded as separate pulses. Pulse pileup occurs when too many photons hit the detector, leading to electronics saturation and count losses. The pulse pileup depends on the photon count rate, the dead time and the effective response time of electronics. The losses due to dead time can become particularly severe at high count rates and the performance of the detector becomes non-linear. The shorter the dead time, the smaller the losses due to dead time.

Assessment of the gamma camera performance requires an understanding of the sources of noise associated with the detection task. It is required to acquire a sufficient number of counts to achieve a statistically significant difference between the target tissue volume and the background count rates. The noise corresponds to unavoidable random fluctuations in the count rates. An increase or decrease in high count rates may be miscalculated leading to a false-positive or a false-negative response, respectively. Significant sources of noise (Barber et al. 1991, Zanzonico and Heller 2000) include the following:

1. Poisson distribution noise, which is related to the counting statistics and is due to the stochastic nature of the counting process. Increasing the counting time or the counting efficiency can lead to a reduction of this type of noise.

2. Noise due to the spatial variations in the adjacent background count rates. These count rate differences related to background sources are a source of random noise always present due to radiation coming from a relatively large volume. A camera with good energy resolution is favoured if this type of noise is dominant because a selection of an appropriate energy window will effectively eliminate background variations due to scattered photons.

In summary, a SFOV nuclear camera should be designed, and its performance operational characteristics should be assessed and optimised on the basis of the demands of the specific imaging task.

2.4 SCINTILLATOR DETECTOR CONFIGURATIONS OF SFOV NUCLEAR CAMERAS

A scintillator-based configuration of a SFOV gamma camera consists of a scintillator crystal optically coupled to a photodetector (PMTs or semiconductor-based photodiodes) and associated readout electronics (Fidler 2000, Levin 2003, Madsen 2007, Moses 2009, Pichler and Ziegler 2004). When incident radiation is absorbed by a scintillator crystal, visible light is generated followed by conversion to an electrical pulse in the photodetector. The charge carriers are optical photons and photoelectrons.

2.4.1 Scintillators

Scintillators, which are transparent luminescent materials that emit visible light when they absorb gamma rays, are the major category of radiation detectors used in nuclear medicine applications (Lecoq 2016, Nassalski et al. 2005, Wilkinson 2004). They may be organic or inorganic solids, liquids or gases. The ones used in nuclear medicine are inorganic crystalline scintillators often with a small amount of impurities that help them scintillate more efficiently. These impurities called activator centres create

TABLE 2.1   Properties of Various Inorganic Scintillator Crystals Used in Small Field of View Gamma Cameras

Scintillator	NaI(Tl)	BGO	LSO:Ce	GSO:Ce	CsI(Tl)	BaF$_2$	Yap:Ce
Density (g/cm³)	3.67	7.13	7.40	6.71	4.51	4.89	5.37
Effective Z	50	74	66	59	54	54	39
Decay time (ns)	230	300	40	60	1000	0.8/620	27
Photon yield (ph/keV)	38	8	25	9	65	11/1.5	18
Refraction index	1.85	2.15	1.82	1.85	1.80	1.56	1.95
Hygroscopic	Yes	No	No	No	No	No	No
Peak emission wavelength (nm)	415	480	420	440	540	220/310	370

Note: BaF$_2$ scintillator has two components: a fast- and a slow-decaying component.

energy levels within the bandgap allowing and enhancing the probability of visible light scintillations during the de-excitation process. The physical properties of the most commonly used scintillation crystals implemented in SFOV gamma camera configurations are listed in Table 2.1.

The scintillator emits visible light proportional to the energy absorbed in the crystal and can be found in two formats: continuous or pixelated crystal array. The segmented crystals offer high intrinsic spatial resolution and significantly fewer edge effects than a continuous crystal; however, crystal segmentation increases the cost and the complexity of the camera and reduces detection sensitivity due to the dead area between the segmented crystals.

The energy band structure and the scintillation process, which is a several-step process, are depicted in Figure 2.2.

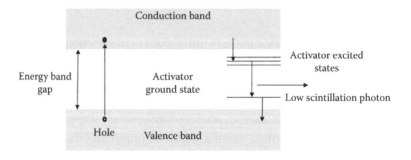

FIGURE 2.2   Energy band structure and scintillation process.

In the ground state, the valence band and the activation centre ground state are filled with electrons, while the conduction and the activation centre excited states are empty. After a γ-ray interaction, electrons in the valence band can absorb energy and get excited into the unfilled conduction band. Since this is not the ground state, the electron de-excites by releasing lower energy scintillation photons and returns to its ground state. These scintillation photons are detected by a photodetector and converted into an electrical signal. The number of scintillation photons produced is proportional to the energy deposited in the crystal. Scintillators require an additional phase, the coupling with a photodetector, to convert the low scintillation light to an electronic signal leading to loss of light, which in turn limits the energy resolution.

The scintillator detectors should have the following fundamental characteristics: high effective atomic number and high density, and therefore large attenuation coefficients, to increase interaction probability and, hence, to increase sensitivity and detection efficiency, short decay time to allow counts to be distinguished at high count rates with minimum dead time and for good coincidence timing and high light output to convert the energy absorbed into detectable light with high efficiency and to have good energy resolution; it also allows a large number of crystal elements per photodetector for high spatial resolution and high conversion efficiency, which is the fraction of deposited energy that is converted into light. Further, they should be transparent to the emitted light to allow efficient light transmission and collection (high intrinsic light yield), and their light yield is proportional to the deposited energy and directly affects the energy resolution. Finally, each scintillator material has a characteristic emission spectrum, and the scintillator emission wavelength should be matched to the properties and the sensitivity of the photodetector.

Advantages of scintillator crystals include reliability, relatively low cost, maturity in the production technology with consistent high quality, high intrinsic counting efficiency and high sensitivity, in particular for medium- and high-energy photons and for high density with high effective Z crystals. Disadvantages include large and bulky size and poor energy resolution leading to not so effective scatter rejection.

2.4.2 Photomultiplier Tube

PMTs are the oldest and most reliable photodetectors which detect low light level scintillation photons and convert them into an electrical signal

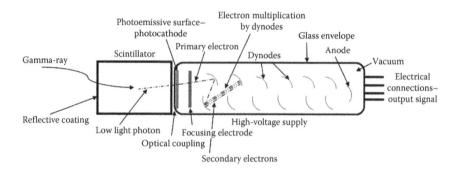

FIGURE 2.3   A schematic diagram of a crystal-photomultiplier tube photo-detector.

(Fidler 2000, Madsen 2007, Moses 2009, Peterson and Furenlid 2011). They have very high electronic gain (10^6) with relatively low noise and fast response. However, they have low quantum conversion efficiency leading to a significant loss of signal that affects both energy resolution and intrinsic spatial resolution. A schematic diagram of a PMT photodetector is shown in Figure 2.3.

They consist of a vacuum enclosure with a very thin photocathode layer at the entrance window. An incoming scintillation photon deposits its energy at the photocathode and releases a photoelectron. Depending on its energy, the photoelectron can escape the surface potential of the photocathode and in the presence of an applied electric field can accelerate to reach the first dynode, a metal plate, which is at a positive potential with respect to the photocathode. A focusing grid provides the electric field required to direct the photoelectrons towards the dynode. When the photoelectron with its increased energy strikes the dynode, it leads to the emission of multiple secondary electrons. This cascade process of acceleration and emission is repeated through a series of several dynodes (9–12), which are at increasing potentials, leading to a gain of more than a million at the final dynode, which is called the anode and sends an output signal to an external circuit. The dynodes are curved to direct the emitted electrons to the next dynode. The absorption of 1 electron at a dynode generates 3–6 electrons depending on the accelerating voltage difference from dynode to dynode. The electron multiplication factor depends on the energy of the photoelectron and the voltage difference between the photocathode and the dynode. The anode can be in the form of a rod, plate or mesh. Electrical connections are made through pins at the end of the tube. The quantum efficiency of a PMT is defined as the ratio of the number of photoelectrons emitted to the number of incident photons and is typically 25%. This low efficiency in the emission and escape of photoelectrons from the cathode is the main drawback of a PMT. However, the high gain obtained by a PMT leads to a very good signal-to-noise ratio for low light levels, which is the primary reason for the success and applicability of PMTs with scintillators. PMTs require a high-voltage supply, which should be very stable because the multiplication factor is very sensitive to dynode voltage changes. The PMT must also be shielded against external magnetic fields to avoid the trajectory of electrons being altered. PMTs come in various sizes and shapes (round, square and hexagonal). When a scintillator is coupled to a PMT, an optical coupling material is added between the two components to minimise reflection losses, while the scintillator is usually surrounded on all other sides by a highly reflective coating material. The most successful detection mode is the one-to-one coupling between a scintillator crystal and a PMT (1 scintillator crystal per PMT). A more recent way of coupling is based on block detector design (Casey and Nutt 1986), which reduces the number of PMTs required for read out by allowing small detector elements (an array: segmented scintillator crystal with saw cuts) to be used and, therefore, improving the spatial resolution while reducing both the complexity and the cost. Four rectangular PMTs are coupled with a segmented scintillator crystal leading to light sharing between the four PMTs. Ratios are formed by combining the signals from all PMTs to provide the spatial coordinates (axial and transverse) of a particular detector element and, therefore, to localise the crystal in which the incident photon interacted. The standard PMT is too bulky for a compact and light SFOV camera.

Different and complex arrangements of the dynode chain have been developed over the years in order to maximise the gain. In particular, a fine mesh grid dynode structure and a cross-wire anode configuration have been implemented to restrict the spread of photoelectrons providing a position-sensitive energy measurement within a single PMT, and this type of PMT is called position sensitive PMT (PSPMT). The PSPMTs offer high gain, high intrinsic spatial resolution, high light distribution, low noise, charge division readout and wide dynamic range. Hence, the PSPMT-based SFOV gamma cameras, with a single or multiple small PSPMTs with positioning capabilities, have several distinct advantages compared to conventional PMT-based cameras, such as smaller size, reduced electronics and lower weight. More recently, a multichannel dynode structure has been developed, which incorporates a microchannel plate and provides ultra-fast response, high sensitivity and low light level detection at photon counting level. This type of PMT is called multichannel PMT, where a multi-anode structure (segmented anode) is used for electron collection providing an extremely improved position-sensitive energy measurement with very little crosstalk between adjacent channels.

2.4.3 Semiconductor-Based Photodetectors

The implementation of semiconductor-based photodetectors to replace the bulky PMTs has been an interesting, notable and timely development (Lewellen 2008, 2010, Pichler and Ziegler 2004, Renker and Lorenz 2009). The conventional photodiodes and the avalanche photodiodes have very high detection efficiency and intrinsic spatial resolution providing a reliable and robust performance compared to PMTs.

A silicon photodiode (SPD) is a semiconductor device which converts light into electrical current. Its operation is based on the same principal as gas-filled detectors, that is, generation of electron–hole pairs in a semiconductor material but with no internal amplification. It consists of a p–n junction, which operates under a reverse voltage bias, that is, a voltage is applied with such a polarity that essentially no electrical current flows. Under reverse bias, a depletion layer is formed which consists of a space charge situated between the p and n regions within the crystal matrix. Operating the photodiode under reverse bias increases the sensitivity as it increases the depletion region, which is the active area of the detector. A SPD's response is linear and the response linearity improves with increasing applied reverse bias. When a light photon enters the SPD, electron–hole pairs are generated in the depletion region due to photon absorption. The electrons and holes are drifted in opposite directions under an externally applied electric field, and a small current, the photocurrent, is generated, which is proportional to the intensity of scintillation light and can be measured by external readout circuits. The SPDs are robust, small, cheap and insensitive to magnetic fields and have high quantum efficiency and small signal-to-noise ratio. However, they have a gain of unity requiring low-noise pre-amplifiers and cooling is required to reduce the leakage current due to thermal noise.

An avalanche photodiode (APD) is also based on a p–n junction diode (Figure 2.4) but is operated at a relatively high reverse bias (typically 100–200 V in silicon), but below the breakdown bias where the Geiger mode

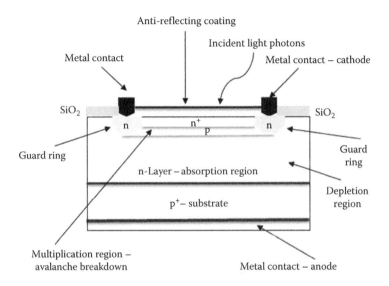

FIGURE 2.4 Schematic diagram of an avalanche photodiode photodetector.

would take over, to enable internal avalanche amplification, which starts a chain of impact ionisation events.

They operate at a lower voltage than PMTs and have a much higher quantum conversion efficiency. When light photons enter the APD, electron–hole pairs are generated. The high reverse bias increases the electric field which in turn increases the velocity and the kinetic energy of electrons in the depletion region to the extent that collisions with the crystal lattice generate secondary electron–hole pairs and some of them can cause further impact ionisation and so on. The net result is an amplification (gain) to the electrical signal and the output photocurrent signal is proportional to the initial number of light photons. Due to the avalanche process, their response is non-linear and suffers from large excess noise. The APDs, either in a single or in an array format, are very sensitive and have high quantum efficiency, fast response and higher gain than the conventional SPDs; however, the gain is very sensitive to small temperature variations and to changes of bias voltage. APDs can operate in a magnetic field.

The silicon photomultipliers (SiPMs) are solid-state devices, which can detect and quantify low light level signals with a single photon sensitivity offering a highly attractive alternative to the low light detection capabilities of a PMT. A SiPM is a densely packed matrix/array of many hundreds or thousands microcell (PD element) APDs on a silicon

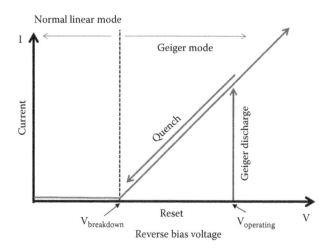

FIGURE 2.5 Geiger-mode operation chain of a silicon photomultiplier photodetector.

substrate operating in Geiger mode. Each microcell, typically between 100–1000 per mm², acts as an independent Geiger-mode detector, which detects photons independently and identically in such a way that when a photon interacts in the microcell, it discharges. The optimal number of microcells depends on the specific application. In Geiger mode, the APD is reverse biased above its breakdown voltage to achieve a very high gain. The Geiger-mode operation consists of a chain of breakdown discharge, quenching and recharge processes and is depicted by the diagram in Figure 2.5.

Electrons and holes will multiply by a cascade of impact ionisations throughout the silicon volume at a faster rate than they can be collected so that the silicon will break down and become conductive leading to exponential growth of current. This process is called Geiger discharge. Once a current flows, it should be stopped or quenched. Each cell is connected to reverse bias voltage by a series resistor resulting in the cell discharge being quenched and, hence, reducing the reverse voltage to a value below its breakdown voltage (recharge/reset phase). The independently operating APD microcells are all connected to a common readout line. The signals of all microcells are summed to form the output of the SiPM. A SiPM produces the same size and shape of a current pulse from each cell regardless of the number of photons interacting within a photodiode at the same time, and therefore, it cannot provide proportional information.

TABLE 2.2   Fundamental Characteristics of Photodetectors: A Comparison

Photodetectors	PMT	APD	SiPM
Active area	1–2000 cm^3	1–100 mm^2	1–10 mm^2
Gain	10^5–10^7	10^2–10^3	10^5–10^7
Excess noise factor	0.1–0.2	>2	1.1–1.2
Quantum efficiency	25%	60%–80%	<40%
Bias voltage (V)	1000–2000	~200–2000	~30–150
Magnetic susceptibility	Very high	No (up to 9.4 T)	No (up to 15 T)

Important advantages of SiPMs include high intrinsic gain with very low operating voltage, robustness, very low excess noise at the single photon level, excellent timing resolution and MR compatibility. However, they have lower photon detection efficiency than APDs, but very similar to the one of PMTs and high dark current count rate which requires cooling. Good linearity requires a large number of microcells to avoid saturation due to multiple interactions in the same microcell. However, the number of microcells should not be too large, because this will lead to an increase in dead space area, which in turn will reduce the mean quantum efficiency over the total area of the device.

A comparison of fundamental characteristics of all the photodetectors, that is, PMTs, APDs and SiPMs, is presented in Table 2.2.

2.5 SEMICONDUCTOR DETECTOR CONFIGURATIONS OF SFOV NUCLEAR CAMERAS

A semiconductor-based configuration of a SFOV gamma camera consists of a compound semiconductor detector and its associated readout electronics. The most commonly used semiconductors in a SFOV camera are cadmium telluride (CdTe), cadmium zinc telluride (CdZnTe) and mercuric iodide (HgI$_2$). Intrinsic properties of the most commonly used semiconductors implemented in SFOV camera configurations are listed in Table 2.3. These semiconductor materials have high Z and density, high stopping power, extremely good energy resolution with no cooling, high intrinsic spatial resolution which leads to changes in the conventional collimator designs, low electron–hole creation energy (W) and high resistivity with low dark current and operate at low voltage. In semiconductor detectors, the gamma-ray energy is directly converted into electron–hole pairs without the need of an intermediate high-gain amplification stage, which produce electrical signals with high energy

TABLE 2.3   Intrinsic Properties of Semiconductor Materials Used in Small Field of View Gamma Cameras

Semiconductor	CdTe	CZT	HgI$_2$	Si
Effective atomic number	48–52	30–52	80–53	14
Density (g/cm^3)	6.06	5.8	6.4	2.33
Electron mobility (cm^2/V/s)	1000	1350	100	1500
Hole mobility (cm^2/V/s)	80	120	4	600
Average energy per electron–hole pair – W (eV)	4.43	4.64	4.15	3.62
Bandgap – E$_g$(eV)	1.47	1.57	2.13	1.12
Resistivity (Ω-cm)	3.0×10^9	3.0×10^{11}	10^{13}	2.3×10^5

resolution (Darambara and Todd-Pokropek 2002, Eisen et al. 2004, Scheiber 2000, Wagenaar 2004).

Over the last decade, due to progress and advances concerning crystal growth techniques and control, electrode pattern configurations, novel metal contact types, improved surface preparation and material uniformity, interconnect technology, ASIC readout electronics, charge loss correction techniques and acceptable spectral performance in terms of energy resolution and spectral stability and relatively low-cost production, CdZnTe (CZT) is considered a very attractive and promising semiconductor material for a radiation detector, and its feasibility has been demonstrated in many spectroscopic and imaging applications (Guerra et al. 2009, Myronakis and Darambara 2011, Szeles 2004, Szeles et al. 2002, Takahashi and Watanabe 2001). CZT is a compound semiconductor with a wide bandgap, high effective atomic number, high density, high stopping power and high resistivity which leads to low leakage current and low-noise characteristics and operates at room temperature without the need of cooling devices to maintain nominal operating conditions.

The main advantage of CZT over scintillators is its unique ability to directly convert the incident photon energy to measured electrical signal (Figure 2.6).

Photons interact with CZT and electron–hole pairs are generated within the charge-free depletion region and drift in opposite directions towards the detector electrode under the influence of an externally applied electric field. Hence, a current signal is induced at the detector collection electrodes with an amplitude proportional to the absorbed energy, which can be measured and further processed externally by the readout electronics.

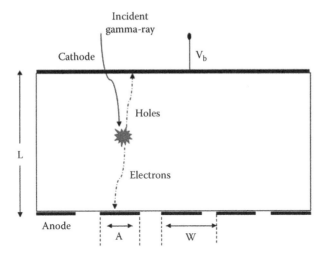

FIGURE 2.6 Schematic diagram of a cadmium zinc telluride semiconductor detector (V_b is the bias voltage, L the material thickness, W the pixel size and A the anode).

In contrast, scintillators require more than one media and stages to produce a measured signal. The entire signal generation process requires at least an order of magnitude more energy to create one carrier pair than in CZT, increasing the statistical noise to the measured signal that degrades the energy resolution.

Taking advantage of some unique properties of CZT detectors, such as increased sensitivity, high intrinsic spatial resolution and excellent energy resolution, CZT detectors may lead to a transformation in nuclear medicine practice offering a new generation of imaging cameras with dose reduction and 3D fast dynamic acquisitions. As a result, there are several CZT-based organ-specific dedicated cameras that have been developed for nuclear cardiology and molecular breast imaging and very SFOV intraoperative handheld miniature probes for radioguided surgery.

Use of CZT detectors to fabricate compact, portable, lightweight and small systems (Figure 2.7) is a very attractive feature, (in comparison with other detectors such as gas filled and scintillators), to produce γ-ray detectors for SFOV gamma cameras.

Advantages of CZT detectors include high detection efficiency and increased sensitivity; high intrinsic spatial resolution (sub-mm) with depth-of-interaction information obtained by the fine pixelated electrode structure, on the front and in the depth, rather than cutting crystals,

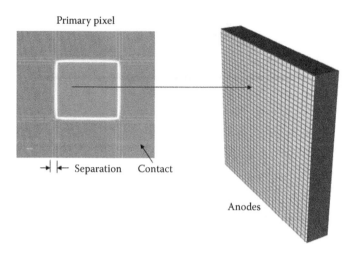

FIGURE 2.7 A cadmium zinc telluride semiconductor detector with pixelated anodes and a primary collection pixel.

as in scintillators; excellent energy resolution, much better than that of scintillators (e.g. CZT, 2% FWHM at 511 keV, while LSO, 15% FWHM at 511 keV) leading to better scatter and random rejection/correction reducing the adverse effects of scattered photons on image quality and for multiple-isotope imaging with reduction of crosstalk; compatibility with MRI, that is, they operate normally with MRI acquisitions up to at least 9.4 T; and detector configurations that are versatile, flexible, lightweight and configurable.

However, CZT detectors have some drawbacks/technical challenges, which are the subject of current ongoing research, such as incomplete charge collection, which limits the detector thickness; lack of material uniformity, which limits charge collection and useful detector size and poor reproducibility, which limits the yield of good detectors and increases costs, in particular for thick crystals and low mobility of charge carriers leading to poor timing performance.

2.6 SFOV BETA CAMERAS

SFOV beta cameras directly image beta radiation with high spatial resolution and sensitivity. Detectors employed for the SFOV beta cameras (Gonzalez et al. 2011, Hoffman et al. 1997, MacDonald et al. 1995, Piert et al. 2007, Strong et al. 2008, 2009) consist of a bundle of plastic scintillators optically coupled via flexible fibre optics to a compact PSPMT. Because of beta's relatively short range in tissues, the contamination from distant sources is significantly reduced and the entire detected beta activity comes from a source very close to the tissue surface that is almost in contact with the camera. As a result, there is no need for high degree of collimation or any other shielding leading to smaller cameras in diameters which are more appealing for intraoperative usage. The beta cameras have high sensitivity to local radioactivity and are mainly used for surface lesions and are implemented in autoradiographic and guided surgical applications.

2.7 FUTURE DEVELOPMENTS

At the time of writing, few clinical studies (Barber et al. 1991, Classe et al. 2005, Hruska et al. 2005, Mariani et al. 2005, Russo et al. 2009) have been performed which directly compare the performance of SFOV gamma cameras based on scintillator and semiconductor configurations; however, they have not provided a clear camera of choice. It is therefore very difficult to evaluate if there are any significant differences in the performance and in the detector capabilities of SFOV nuclear cameras in clinical settings. Ongoing and future improvements and developments in SFOV gamma camera detector technologies and configurations are necessary to optimise the overall performance of the SFOV gamma cameras and to improve sensitivity, specificity, spatial resolution, flexibility and operator friendliness. In addition, SFOV gamma cameras should continue being small and compact, but they should be able to image the entire or most of the surgical field in a single view. In the long run, the semiconductor-based photodetectors will replace the PMTs. Further advances in CZT detectors will make them a more attractive possibility for being implemented into SFOV gamma camera configurations and intraoperative imaging applications. New PET probes and optical imaging will also play a greater role in SFOV cameras. Finally, future developments in SFOV nuclear cameras should include 3D volumetric imaging and multimodal configurations (Beijst et al. 2016, Lees et al. 2014) to improve localisation and procedure guidance as well as innovative approaches, which will include dynamic integration of surgery, surgical robotics and real-time imaging for improved accuracy and precision to enable better decision making and integration of therapy with intraoperative imaging for enhanced diagnostic accuracy to enable timely and precise treatment.

REFERENCES

Abe, A., N. Takahashi, J. Lee et al. 2003. Performance evaluation of a hand-held, semiconductor (CdZnTe)-based gamma camera. *European Journal of Nuclear Medicine and Molecular Imaging*, 30(6):805–811.

Anger, H.O. 1958. Scintillation camera. *Review of Scientific Instruments*, 29:27–33.

Barber, H.B., H.H. Barrett, T.S. Hickernell et al. 1991. Comparison of NaI (Tl), CdTe, and HgI2 surgical probes: Physical characterization. *Medical Physics*, 18(3):373–381.

Beijst, C., M. Elschot, M.A. Vierghevere and H.W.A.M. de Jong. 2016. Toward simultaneous real-time fluoroscopic and nuclear imaging in the intervention room. *Radiology*, 278(1):232–238.

Bhatia, B.S., S.L. Bugby, J.E. Lees and A.C. Perkins. 2015. A scheme for assessing the performance characteristics of small field-of-view gamma cameras. *Physica Medica*, 31(1):98–103.

Bugby, S.L., J.E. Lees, B.S. Bhatia and A.C. Perkins. 2014. Characterisation of a high resolution small field of view portable gamma camera. *Physica Medica*, 30(3):331–339.

Casey, M.E. and R. Nutt. 1986. A multicrystal two-dimensional BGO detector system for positron emission tomography. *IEEE Transaction on Nuclear Science*, 33:460–463.

Cherry, S.R., J.A. Sorenson and M.E. Phelps. 2003. *Physics in Nuclear Medicine*, 3rd edn. Philadelphia, PA: W. B. Saunters.

Classe, J.M., M. Fiche, C. Rousseau et al. 2005. Prospective comparison of 3 gamma-probes for sentinel lymph node detection in 200 breast cancer patients. *Journal of Nuclear Medicine*, 46(3):395–399.

Darambara, D.G. and A. Todd-Pokropek. 2002. Solid state detectors in nuclear medicine. *Quarterly Journal of Nuclear Medicine and Molecular Imaging*, 46(1):3.

Eisen, Y., A. Shor and I. Mardor. 2004. CdTe and CdZnTe X-ray and gamma-ray detectors for imaging systems. *IEEE Transaction on Nuclear Science*, 51(3):1191–1198.

Fidler, V. 2000. Current trends in nuclear instrumentation in diagnostic nuclear medicine. *Radiology and Oncology*, 34(4):381–385.

Fiorini, C., A. Longoni, F. Perotti et al. 1999. First prototype of a gamma-camera based on a single CsI(Tl) scintillator coupled to a silicon drift detector array. In *IEEE Nuclear Science Symposium and Medical Imaging Conference Record*, Seattle, WA.

Goertzen, A.L., J.D. Thiessen, B. McIntosh, M.J. Simpson and J. Schellenberg. October 2013. Characterization of a handheld gamma camera for intraoperative use for sentinel lymph node biopsy. In *IEEE Nuclear Science Symposium and Medical Imaging Conference Record*, Seoul, Korea, pp. 1–4.

González, S.J., L. González, J. Wong et al. 2011. An analysis of the utility of hand-held PET probes for the intraoperative localization of malignant tissue. *Journal of Gastrointestinal Surgery*, 15(2):358–366.

Guerra, P., A. Santos and D.G. Darambara. 2009. An investigation of performance characteristics of a pixellated room-temperature semiconductor detector for medical applications. *Journal of Physics D: Applied Physics*, 42:175101.

Halkar, R.K. and J.N. Aarsvold. 1999. Intraoperative probes. *Journal of Nuclear Medicine Technology*, 27(3):188–193.

Hartsough, N.E., H.H. Barrett, H.B. Barber and J.M. Woolfenden. 1995. Intraoperative tumor detection: Relative performance of single-element, dual-element and imaging probes with various scintillators. *IEEE Transactions on Medical Imaging*, 14(2):259–265.

Heller, S. and P. Zanzonico. May 2011. Nuclear probes and intraoperative gamma cameras. *Seminars in Nuclear Medicine*, 41(3):166–181.

Hoffman, E.J., M.P. Tornai, M. Janecek, B.E. Patt and J.S. Iwanczyk. 1999. Intraoperative probes and imaging probes. *European Journal of Nuclear Medicine*, 26(8):913–935.

Hoffman, E.J., M.P. Tornai, M. Janecek, B.E. Patt and J.S. Iwanczyk. 2004. Intraoperative probes and imaging probes. In *Emission Tomography: The Fundamentals of PET and SPECT*, eds. M.N. Wernick and J.N. Aarsvold, pp. 335–353. London, UK: Elsevier Academic Press.

Hoffman, E.J., M.P. Tornai, S.C. Levin, L.R. MacDonald and S. Siegel. 1997. Gamma and beta intraoperative imaging probes. *Nuclear Instruments and Methods in Physics Research A: Accelerators, Detectors and Associated Equipment*, 392:324–329.

Hruska, C.B., M.K. O'Connor and D.A. Collins. 2005. Comparison of small field of view gamma camera systems for scintimammography. *Nuclear Medicine Communications*, 26(5):441–445,

Keshtgar, M.R.S. and P.J. Ell. 1999. Sentinel lymph node detection and imaging. *European Journal of Nuclear Medicine and Molecular Imaging*, 26(1):57–67.

Kim, J.H., Y. Choi, K.S. Joo et al. 2000. Development of a miniature scintillation camera using an NaI (Tl) scintillator and PSPMT for scintimammography. *Physics in Medicine and Biology*, 45(11):3481.

Knoll, G.F. 2000. *Radiation Detection and Measurement*, 3rd edn. New York: John Wiley & Sons.

Knoll, P., S. Mirzaei, K. Schwenkenbecher and T. Barthel. 2014. Performance evaluation of a solid-state detector based handheld gamma camera system. *Frontiers in Biomedical Technologies*, 1(1):61–67.

Kopelman, D., I. Blevis, G. Iosilevsky et al. 2005. A newly developed intra-operative gamma camera: Performance characteristics in a laboratory phantom study. *European Journal of Nuclear Medicine and Molecular Imaging*, 32(10):1217–1224.

Kwo, D.P., H.B. Barber, H.H. Barrett, T.S. Hickernell and J.M. Woolfenden. 1991. Comparison of NaI (Tl), CdTe, and HgI2 surgical probes: Effect of scatter compensation on probe performance. *Medical Physics*, 18(3):382–389.

Lecoq, P. 2016. Development of new scintillators for medical applications. *Nuclear Instruments and Methods in Physics Research Section A: Accelerators, Detectors and Associated Equipment*, 809:130–139.

Lees, J.E., D.J. Bassford, O.E. Blake, P.E. Blackshaw and A.C. Perkins. 2011. A high resolution Small Field Of View (SFOV) gamma camera: A columnar scintillator coated CCD imager for medical applications. *Journal of Instrumentation*, 6(12):C12033.

Lees, J.E., S.L. Bugby, B.S. Bhatia et al. 2014. A small field of view camera for hybrid gamma and optical imaging. *Journal of Instrumentation*, 9:C12020.

Levin, C.S. 2003. Detector design issues for compact nuclear emission cameras dedicated to breast imaging. *Nuclear Instruments and Methods in Physics Research Section A: Accelerators, Spectrometers, Detectors and Associated Equipment*, 497(1):60–74.

Levin, C.S. 2004. Application-specific small field-of-view nuclear emission imagers in medicine. In *Emission Tomography: The Fundamentals of PET and SPECT*, eds. M.N. Wernick and J.N. Aarsvold, pp. 293–328. London, UK: Elsevier Academic Press.

Levin, C.S., E.J. Hoffman, M.P. Tornai and L.R. MacDonald. 1996. PSPMT and PIN diode designs of a small scintillation camera for imaging malignant breast tumors. In *IEEE Nuclear Science Symposium Conference Record*, Anaheim, CA.

Lewellen, T.K. 2008. Recent developments in PET detector technology. *Physics in Medicine and Biology*, 53(17):R287.

Lewellen, T.K. 2010. The challenge of detector designs for PET. *American Journal of Roentgenology*, 195(2):301–309.

MacDonald, L.R., M.P. Tornai, C.S. Levin et al. 1995. Investigation of the physical aspects of beta imaging probes using scintillating fibers and visible light photon counters. *IEEE Transactions on Nuclear Science*, 42(4):1351–1357.

Madsen, M.T. 2007. Recent advances in SPECT imaging. *Journal of Nuclear Medicine*, 48(4):661–673.

Mariani, G., A. Vaiano, O. Nibale and D. Rubello. 2005. Is the "Ideal" γ-probe for intraoperative radioguided surgery conceivable? *Journal of Nuclear Medicine*, 46:388–390.

Matthews II, K.L. 1997. Development and application of a small gamma camera. *Medical Physics*, 24(11):1802–1802.

Moses, W.W. 2009. Photodetectors for nuclear medical imaging. *Nuclear Instruments and Methods in Physics Research Section A: Accelerators, Spectrometers, Detectors and Associated Equipment*, 610:11–15.

Mueller, B., M.K. O'Connor, I. Blevis et al. 2003. Evaluation of a small CDZnTe detector for scintimammography. *Journal of Nuclear Medicine* 44:602–609.

Myronakis, M.E. and D.G. Darambara. 2011. Monte Carlo investigation of charge-transport effects on energy resolution and detection efficiency of pixelated CZT detectors for SPOECT/PET applications. *Medical Physics*, 38(1):455–467.

Nassalski, A., M. Kapusta, T. Batsch et al. 2005. Comparative study of scintillators for PET/CT detectors. In *IEEE Nuclear Science Symposium Conference Record*, Puerto Rico, pp. 2823–2829.

Olcott, P., G. Pratx, D. Johnson, E. Mittra, R. Niederkohr and C.S. Levin. 2014. Clinical evaluation of a novel intraoperative handheld gamma camera for sentinel lymph node biopsy. *Physica Medica*, *30*(3):340–345.

Olcott, P.D., F. Habte, A. Foudray et al. 2007. Performance characterization of a miniature high sensitivity gamma ray camera. *IEEE Transactions on Nuclear Science*, *54*:1492–1497.

Parnham, K.B., J. Grosholz, R.K. Davis, S. Vydrin and C.A. Cupec. 2001. Development of a CdZnTe-based small field-of-view gamma camera. SPIE Proceedings: Penetrating Radiation Systems & Applications III, 4508: 134–140.

Patt, B.E., M.P. Tornai, J.S. Iwanczyk, C.S. Levin and E.J. Hoffman. 1997. Development of an intraoperative gamma camera based on a 256-pixel mercuric iodide detector array. *IEEE Transactions on Nuclear Science*, 44(3):1242–1248.

Peterson, T.E. and L.R. Furenlid. 2011. SPECT detectors: The Anger camera and beyond. *Physics in Medicine and Biology*, *56*(17): R145–R182.

Pichler, B.J. and S.I. Ziegler. 2004. Photodetectors. In *Emission Tomography: The fundamentals of PET and SPECT*, eds. M.N. Wernick and J.N. Aarsvold, pp. 255–270. London, UK: Elsevier Academic Press.

Piert, M., M. Burian, G. Meisetschlager et al. 2007. Positron detection for the intraoperative localisation of cancer deposits. *European Journal of Nuclear Medicine and Molecular Imaging*, *34*:1534–1544.

Povoski, S.P., R.L. Neff, C.M. Mojzisik et al. 2009. A comprehensive overview of radioguided surgery using gamma detection probe technology. *World Journal of Surgical Oncology*, *7*:11.

Raylman, R.R. 2001. Performance of a dual, solid-state intraoperative probe system with 18F, 99mTc, and 111In. *Journal of Nuclear Medicine*, *42*(2): 352–360.

Raylman, R.R. and R.L. Wahl. 1994. A fiber-optically coupled positron-sensitive surgical probe. *Journal of Nuclear Medicine: Official Publication, Society of Nuclear Medicine*, *35*(5):909–913.

Renker, D. and E. Lorenz, E. 2009. Advances in solid state photon detectors. *Journal of Instrumentation*, *4*(04):P04004.

Russo, P., G. Mettivier, R. Pani, R. Pellegrini, M.N. Cinti and P. Bennati. 2009. Imaging performance comparison between a LaBr3: Ce scintillator based and a CdTe semiconductor based photon counting compact gamma camera. *Medical Physics*, *36*(4):1298–1317.

Sanchez, F., J.M. Benlloch, B. Escat et al. 2004. Design and tests of a portable mini gamma camera. *Medical Physics*, *31*(6):1384–1397.

Sarikaya, I., A. Sarikaya and R.C. Reba. November 2008. Gamma probes and their use in tumor detection in colorectal cancer. *International Seminars in Surgical Oncology*, *5*(1):25.

Sauret, S.A.B.V., P. Bernhardt, B. Wangberg, H. Ahlman and E. Forsell-Aronsson. 1999. 111In-Labeled radiopharmaceuticals. *Journal of Nuclear Medicine*, *40*:2094–2101.

Scheiber, C. 2000. CdTe and CdZnTe detectors in nuclear medicine. *Nuclear Instruments and Methods in Physics Research Section A: Accelerators, Spectrometers, Detectors and Associated Equipment,* 448(3):513–524.

Schneebaum, S., E. Even-Sapir, M. Cohen et al. 1999. Clinical applications of gamma-detection probes–radioguided surgery. *European Journal of Nuclear Medicine,* 26(1):S26–S35.

Strong, V.E., C.J. Galanis, C.C. Riedel et al. 2009. Portable PET probes are a novel tool for intraoperative localization of tumor deposits. *Annals of Surgical Innovation and Research,* 3:2–8.

Strong, V.E., J. Humm, P. Russo et al. 2008. A novel method to localize antibody-targeted cancer deposits intraoperatively using handheld PET beta and gamma probes. *Surgical Endoscopy,* 22(2):386–391.

Szeles, C. 2004. Advances in the crystal growth and device fabrication technology of CdZnTe room temperature radiation detectors. *IEEE Transactions on Nuclear Science,* 51(3):1242–1249.

Szeles, C., S.E. Cameron, J.O. Ndap and W.C. Chalmers. 2002. Advances in the crystal growth of semi-insulating CdZnTe for radiation detector applications. *IEEE Transaction on Nuclear Science,* 49(5):2535–2540.

Takahashi, T. and S. Watanabe. 2001. Recent progress in CdTe CdZnTe detectors. *IEEE Transactions on Nuclear Science,* 48:950–959.

Tsuchimochi, M. and K. Hayama. 2013. Intraoperative gamma cameras for radioguided surgery: Technical characteristics, performance parameters, and clinical applications. *Physica Medica,* 29(2):126–138.

Tsuchimochi, M., K. Hayama, T. Oda, M. Togashi and H. Sakahara. 2008. Evaluation of the efficacy of a small CdTe γ-Camera for sentinel lymph node biopsy. *Journal of Nuclear Medicine,* 49(6):956–962.

Tsuchimochi, M., H. Sakahara, K. Hayama et al. 2003. A prototype small CdTe gamma camera for radioguided surgery and other imaging applications. *European Journal of Nuclear Medicine and Molecular Imaging,* 30(12):1605–1614.

Valdés Olmos, R.A., S. Vidal-Sicart and O.E. Nieweg. 2010. Technological innovation in the sentinel node procedure: Towards 3-D intraoperative imaging. *European Journal of Nuclear Medicine and Molecular Imaging,* 37(8):1449–1451.

Wagenaar, D.J. 2004. CdTe and CdZnTe semiconductor detectors for nuclear medicine imaging. In *Emission Tomography: The fundamentals of PET and SPECT,* eds. M.N. Wernick and J.N. Aarsvold, pp. 270–284. London, UK: Elsevier Academic Press.

Wilkinson III, F. 2004. Scintillators. In *Emission Tomography: The Fundamentals of PET and SPECT,* eds. M.N. Wernick and J.N. Aarsvold, pp. 229–255. London, UK: Elsevier Academic Press.

Williams, M.B., A.R. Goode, V. Galbis-Reig., S. Majewski, A.G. Weisenberger and R. Wojcik. 2000. Performance of a PSPMT based detector for scinti-mammography. *Physics in Medicine and Biology,* 45(3):781.

Zanzonico, P. and S. Heller. January 2000. The intraoperative gamma probe: Basic principles and choices available. *Seminars in Nuclear Medicine,* 30(1):33–48.

CHAPTER 3

Surgical Experience with POCI and TReCam Prototype Cameras

M.-A. Duval, A. Bricou, E. Barranger, K. Kerrou,
S. Pitre, L. Ménard, B. Janvier, F. Lefebvre,
L. Pinot, M.-A. Verdier and Y. Charon

CONTENTS

3.1 INTRODUCTION

In the field of the surgical treatment of cancer, radiation-counting detectors are used in the operating theatre to help the surgeon localise and remove the radiolabelled tissues. This radioguiding technique allows the precise localisation of the tissues and therefore aids their reliable complete excision. Like a mine-clearing expert using a metal detector, the surgeon is informed by a beep getting sharper and more frequent as the probe gets closer to the radioactive focus. Some radioactive areas have been missed when the radioactive foci are weak and deeply seated in the tissue (>5 cm) or if an intense adjacent source overwhelms the probe and prevents a small signal being detected.

In order to improve this 'blind' search, mini-gamma cameras have been introduced in operating rooms to complement the counting probes. Many different handheld or arm-mounted cameras have been designed. Their place and usefulness are still debated (Bricou et al. 2013). They have been evaluated in many pathologies, but mainly breast cancer and melanoma.

This chapter presents two portable gamma cameras built in our laboratory, 'imaging and modelling in neurobiology and cancerology', in Orsay, France. We not only have designed the cameras in close collaboration with the nuclear physicians and the surgeons but also have worked with them during the clinical evaluation on breast cancer. First, the scope of the work is discussed and then the two cameras and their associated results are presented.

3.2 FRAME OF WORK

Our cameras have been evaluated on patients suffering from early-stage breast cancer. When the tumour is still small, that is, less than 2 cm in its biggest dimension, patients are eligible for the sentinel node (SN) biopsy. This procedure consists of excising the first lymphatic nodes, draining the tumour and establishing their metastatic status. This staging status reflects the spreading of the cancer cells in the body; therefore, this information is needed to inform the decision process the medical team needs to make in order to elaborate the treatment.

At least one SN must be excised to fulfil the procedure. Depending on the surgeon, up to five SNs can be collected. These nodes must be located through the skin by the surgeon before incision. Although their position depends on the tumour's location in most instances, they are seated in the axilla, but their localisation must be known precisely to avoid morbidity due to a too large incision.*

Finding the nodes draining the breast tumour is achieved by injecting a lymph-draining radiotracer in or next to the tumour or around the nipple. This method is built on the hypothesis that the tracers follow the lymphatic vessels to the nodes which may retain them. In fact the transport of these agents depends on their diameter (Cousins et al. 2014). For the blue dye which is less than 10 nm, there is high diffusion and fast migration through the system. A few minutes after injection in the theatre, blue nodes can be seen as long as there is no tissue covering them. For the 140 keV gamma-emitting technetium-radiolabelled nanoparticles that are colloids of about 50–200 nm, the migration speed is lower, but retention in the nodes is higher as particle size increases. In the hospitals where our studies were performed, the ^{99m}Tc injection was always given the day before the operation, mostly for logistical reasons with surgery being performed early in the morning. Generally, lymphatic drainage can be observed on the preoperative scintigraphic images 1 or 2 h after injection. The injected dose is calculated so that the remaining radioactivity is sufficient for the probe to be used the day after injection. Indeed, ^{99m}Tc has a 6 h decay half-life which must be taken into account by the gamma camera designers along with the delay between the injection and the node hunting. Knowing that only 1% on average of the injected radioactivity goes to the SN (Wendler et al. 2010), one can calculate if the expected count rate for the SN can be detected above the ambient background noise taking into account the camera efficiency.

The injection sites have also been a subject of debate (Bernard et al. 2005). One argument against the injection inside the tumour is to avoid accidentally flushing metastases out of the tumour, therefore spreading the disease while trying to treat it.

An issue of SN biopsy (SNB) is the false-negative rate as it can lead to under-treatment for these patients whose node status is unknown. Therefore, the efforts are focused on how to lower this rate. The reference method is to inject two tracers: a radioactive ^{99m}Tc-labelled product and a blue dye (Aarsvold and Alazraki 2005). This not only does improve the SN detection rate (estimated at 95%) but also reduces the false-negative

---

* One of the benefits of this procedure is that if the nodes are found to be non-metastatic, no axillary dissection is performed avoiding pain, arm swelling and loss of mobility to the patient.

rate; therefore, risk of axillary relapse when axillary dissection is not undertaken because the SN is found to be non-metastatic. Although experts are still debating whether preoperative lymphoscintigraphy (LS) is useful, it is still the standard of care in our case (Schwartz et al. 2002; Kawase et al. 2006).

We have designed and built successively two portable gamma cameras named per-operative compact imager (POCI) and tumour resection camera (TReCam). They both provide real-time images of the planar radioactive tracer distribution. Our idea is to work closely with surgeons in order to meet their needs but also to suggest to them new working practices.

POCI was evaluated on the SNB standard procedure.

3.3 POCI

3.3.1 Technical Aspect

The handheld gamma camera POCI combines a high-resolution imaging head with a round intensified position-sensitive diode (IPSD) that gives the camera its distinctive shape. The camera has an outer diameter of 9.5 cm and a height of 9 cm (Figure 3.1).

POCI weighs 1.2 kg which makes it easy to hold in one hand without much effort. The gamma rays are selected by a high-resolution parallel-hole lead collimator. It has a 15 mm thickness which is half that of standard gamma cameras. This is sufficient as POCI is placed on the patient's skin and detects radioactive foci within a 5 cm frontal volume. The collimator holes are hexagonal and the distance between the two parallel sides of a hole is 1.4 mm. The configuration of the holes is also hexagonal. The septal distance is 0.3 mm. The gamma rays are then converted into visible light in a 3 mm thick CsI(Na) crystal plate. This imaging head covers a 40 mm diameter field of view (FOV; 12.5 cm^2; Pitre et al. 2003).

POCI exhibits the following performances at 140 keV: a spatial resolution of 2.2 mm which is sufficient to detect SN whose biggest dimension is typically 1 cm, a sensitivity of 280 cps/MBq and an energy resolution of 28%.

3.3.2 Study

We believe that small gamma cameras can replace the standard gamma cameras to perform the preoperative LS without affecting patient outcome, and moreover, they can be useful tools for the surgeon since they can be used in the theatre to complement the counting probe. We designed

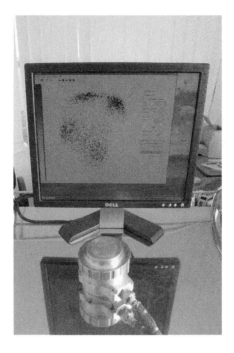

FIGURE 3.1   Per-operative compact imager and one image showing four sentinel nodes.

a prospective and blind study with two objectives: the main goal was to compare the performance of POCI with conventional preoperative LS and the secondary objective was to evaluate the intraoperative performance of POCI to show and quantify the complementarity of this small imager to the probe performance. POCI, if shown not to be inferior to conventional preoperative LS regarding the number of sentinel lymph nodes detected, could be used routinely to create a radioactive map of the axillary or extra-axillary areas.

3.3.3 Protocol

A number of patients (162) with a breast cancer tumour size of less than 20 mm (final average size was 15 mm) were included in the trial in Tenon hospital, Paris, between January 2006 and February 2008. The SNB protocol consists of four periareolar injections of 37 MBq ^{99m}Tc-labelled nanocolloid (Nanocis®, CIS Bio International), 2 h before LS. The first preoperative LS was performed in the nuclear medicine department by a nuclear physician using a standard triple-head gamma camera (IRIX Marconi, Philips)

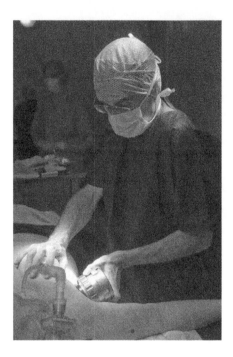

FIGURE 3.2   Looking for sentinel nodes in patient's axilla with per-operative compact imager.

equipped with high-resolution low-energy collimators. Anterior and posterior simultaneous views were acquired in 5 min. SN number and localisations were recorded. After the patient was back in her hospital bedroom, a second LS was achieved with POCI by another nuclear physician blinded to the results of the first LS. The whole axilla was scanned in 5–10 s shots; in addition, a screening for extra-axillary SNs was performed. The day after, the surgeon performed SN scan with POCI before incision recording the axillary SN place and number (Figure 3.2). After removing SNs with the help of the counting probe and the blue dye, the surgeon scanned again the axilla to check for residual hot nodes.

3.3.4 POCI-in-Theatre versus Standard LS Results

We later narrowed the number of patients to 138 who were eligible in meeting the main selection criterion. Detailed results comparing POCI with standard gamma camera images the day before surgery can be found in Kerrou et al. (2011). We present here unpublished results on 131 patients whose data made it possible to compare the performances of POCI in the

TABLE 3.1   Sentinel Node Identification by Classical Lymphoscintigraphy the Day before Surgery and Per-Operative Compact Imager in Theatre Bloc ($N = 131$)

	LS	POCI
Number of axillary nodes (mean; range)	227 (1.7; 0–7)	213 (1.5; 0–4)
Number of extra-axillary nodes (mean; range)	30 (0.2; 1–4)	4 (0.03; 0–1)
Number of patients with no SN identified	6 of 131	8 of 131

Note: SN, sentinel node; LS, lymphoscintigraphy.

theatre (what we call lymphoscintigraphy before surgery) and the standard LS. Eleven patients were excluded a posteriori based on exclusion or non-inclusion criteria: size of tumour superior to 20 mm ($N = 5$), clinically palpable axillary lymph node ($N = 2$), prior history of breast plastic surgery ($N = 2$), multifocal breast cancer ($N = 1$) and benign tumour in one patient. The protocol was not followed for one patient (one exclusion). Due to the lack of operator, 7 patients are missing and 12 more due to maintenance of POCI. Due to these 12 exclusions and 19 non-eligible patients, the results regarding the main judgment criteria deal with 131 patients. The results showed that the LS detected slightly more SNs (227 versus 213; see Table 3.1).

Nevertheless, POCI has a better spatial resolution than the standard gamma camera, and therefore, it should detect more SNs. By studying the discordant cases where the LS results are different from that obtained with POCI, they happen to be systematically cases with multiple SNs seen by the LS. We believe that several reasons may be the cause of the difference in results between the POCI and the LS. One reason is that if an SN's uptake is already small the day before surgery, the signal will be weaker in the theatre due to radioactive decay. We believe that these results are due to the different positions of the patient for the LS exam and for POCI in the operating room. Indeed, for the LS, the patient is in back decubitus position, with hands behind the head, which unwinds the lymph node chains completely. On the operating table, the patient's arm is perpendicular to the body axis so axillary nodes may overlay. In addition to these explanations, we do not neglect the operator-dependent factor. To avoid this bias, the Italian study reported that the same nuclear physician's team used the imaging probe (IP) in the operating room for all patients (Scopinaro et al. 2008). In our study, we chose to work in realistic conditions with a six-surgeon team as operators.

Nevertheless, these good results on a significant number of patients showed that POCI can be an alternative tool to the standard LS.

This aspect is especially interesting in the development of the SNB procedure because the general recommendations require LS in the SNB procedure in breast cancer treatment. This examination highlights some of the limitations in the use of standard gamma cameras, mainly lack of portability and the duration of the procedure which requires the patient to be present in the nuclear medicine department for several hours. The interest of these new portable gamma cameras is to replace the standard gamma cameras which would simplify the general procedure and would even facilitate access to the SNB for patients treated in hospitals with no nuclear medicine department.

Furthermore, if the necessity for LS 1 day before operating is controversial, using POCI in the operating room makes a lot more sense. The surgeon fully adopted POCI and also used it as a probe. First, the surgeon quickly explored the axilla area with POCI so as not to prolong the operation duration and to check if some radioactive SNs were visible. One can say that the operator used POCI as the probe before incision. Thus, in practice, the surgeon confirmed the presence of radioactive SN but did not seek to perform a LS with a complete numbering and an accurate localisation of all SNs. However, POCI was particularly valuable to confirm the biopsy quality by checking the absence of radioactivity in the wound.

3.3.5 POCI as a Complementary Tool to the Probe

In all patients, at least one SN was identified during surgical exploration. A total of 284 SNs were excised (average 2.2 per patient, range 1–6). Among these excised 284 SNs, 272 (96%) were radioactive of which 188 were both blue and radioactive and 84 were radioactive alone. Only a few SNs (12/284 = 4%) were coloured blue but not radioactive.

In the operating room, POCI did not detect any SN in 8 cases out of 131 patients among which 2 of these were in common with the LS. POCI detected 213 axillary SNs (average 1.6 per patient, range 0–4; see Table 3.2).

TABLE 3.2   Sentinel Node Identification by Per-Operative Compact Imager and Counting Probe in the Operating Room ($N = 131$)

	POCI	Counting Probe	Residual SNs	Total
Number of axillary nodes (mean; range)	213 (1.6; 0–4)	251 (2.07; 0–6)	21(0.16; 1–2)	272

Note: SN, sentinel node.

The sensitivity was 78% (213/272, 95% confidence interval 73–83) for POCI. Four extra-axillary SNs (range 0–1) were detected in four patients. More precisely, before incision, POCI detected 198 axillary SNs and 3 extra-axillary SNs. After incision and the detection of 251 axillary SNs by the counting probe, POCI identified 21 more SNs which we called residual SNs in 20 patients (15% of the cases). Among these 21 residual SNs, 6 had already been located and counted by POCI before incision. Among the 272 excised radioactive SNs, the residual ones represent 8%. Without POCI, they would not have been excised. Figure 3.3 shows three SNs clearly detected by POCI.

In our study, the POCI camera detected residual SNs in 15% of the patients. In one case (patient N°33), the detection of one residual SN helped avoid a lymph node dissection to the patient because it was the only SN to be found. It was negative at the histopathology analysis. In another case (patient N°22), POCI detected an additional metastatic SN not identified by the counting probe changing the final adjuvant treatment due to a better staging. Among published literature, only two clinical trials have reported the presence of residual SNs. Mathelin et al. reported one case of

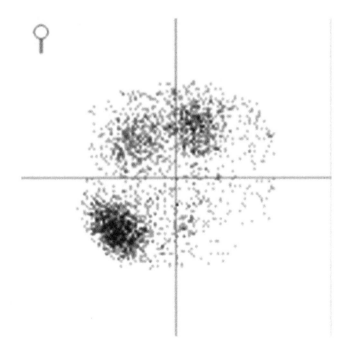

FIGURE 3.3 Lymphoscintigraphy performed by per-operative compact imager. Three sentinel nodes were clearly identified. Image acquisition time was 79 s.

residual SN among 25 patients and Motomura et al. reported three cases from 29 patients detected by the compact gamma camera (Motomura et al. 2005; Mathelin et al. 2008).

We think that imagers like POCI are indeed an excellent tool in addition to the probes, in particular, anatomical and physiological situations. This performance difference between both per-operating systems is based on their intrinsic detection techniques. A probe's geometric efficiency is the solid angle which goes down like the inverse of the square of distance between the probe and the radioactive source. For example, geometric efficiency drops by approximately 90% from 1 to 5 cm in depth. This implies that the probe shows difficulties in finding SNs deeply located in the tissues, whereas POCI's detection efficiency is not degraded by solid angle effect but only very little by attenuation and Compton scattering (240–117 cps/MBq at source-to-detector distances of 1 and 5 cm, respectively).

Moreover, in the case of weakly radioactive SNs, the imager has better detection efficiency due to its ability to discriminate specific signal from noise, thanks to the spatial distribution of the radioactivity detected. This intrinsic property of imaging enhances the signal from noise detection with respect to counting probe. This feature was previously reported in a Japanese study (Motomura et al. 2005).

Using POCI as a complementary tool to the probe increases the detection of multiple nodes in a patient. Scopinaro et al. (2008) reported more detected SN with the IP. The direct clinical impact demonstrated in our study is to decrease the probability of missing one node and to modify the therapeutic strategy of the patient.

3.4 TRECAM

After the POCI trial's success, we built another portable gamma camera based on a different technology. There were two reasons for this decision: the parts in POCI were no longer available, especially the IPSD, and the surgeon asked for a bigger FOV to scan the axilla in one shot. So we doubled the FOV, but as a drawback, we also doubled the weight. The surgeon was also interested to use this camera, named TReCam, in a new protocol in France: the sentinel node and occult lesion localisation (SNOLL; Feggi et al. 2001; Thind et al. 2011). This procedure consists of an SNB coupled with radioguided lesion localisation. Indeed, due to the extension of screening, 25% of breast cancer lesions are non-palpable (in general <0.5 cm). In such cases, the surgeon needs some help to localise it inside the wound so the tumour can be completely resected with cancer-free margins to avoid reoperation. Another issue is the conservation of, as much as possible, the breast tissue since the lesion is very small. One classical way to spot the lesion is to insert a wire hook into it before surgery so the surgeon can track it down. The other way is to inject ^{99m}Tc-colloids in the vicinity of the tumour. Most of the radiotracer stays there and a small quantity (on average 1%) flows to the nodes. So with the same injection, the surgeon can be radioguided to the lesion and the SN: this is the SNOLL protocol. The radioguiding tools are the usual counting probes with an extra collimator for the lumpectomy because it is 100 times more active than the SN. This procedure has been increasingly used with good results (Lovrics et al. 2011; Giacalone et al. 2012). The original idea was to use TReCam in addition to the probe and see if it improved the outcomes. The main criterion is the re-intervention rate in case the margins are not cancer-free.

3.4.1 Technical Aspect and Design of the Detection Head of TReCam

TReCam was designed to be handheld so as to be easily moved and positioned on the patient's skin (axilla armpit, breast). The current prototype measures 117 mm × 114 mm × 83 mm and weighs 2.2 kg. It is composed of a parallel-hole collimator, a LaBr$_3$(Ce) crystal plate and a multi-anode photomultiplier tube (MAPMT). The 256 output signals of the MAPMT are read out by front-end electronics composed of four 64-channel HARDROC2 ASICS (Callier et al. 2007), with individual input, trigger and charge measurement. A 4 mm thick lead shielding surrounds the scintillator and half the height of the MAPMT and stops 99% of the gamma photons that could come from the sides. The prototype is housed in an aluminium box (Figure 3.4) and connected to a computer via a 5 m long wire cable (Figure 3.5). The entire intraoperative probe system was designed according to electromedical safety regulations.

3.4.2 Imaging Head

The imaging head of our prototype combines a parallel-hole collimator with a continuous crystal plate, covering a square FOV of 49 mm × 49 mm. The lead collimator, manufactured by Nuclear Fields, is composed of a 15 mm thick stack with 1.5 mm diameter hexagonal holes and 0.23 mm septum thickness. It leads to a geometrical efficiency of $5 \cdot 10^{-4}$ and a 6.5 mm spatial resolution at 5 cm. The scintillating crystal is a 6 mm thick LaBr$_3$(Ce) plate manufactured by Saint Gobain Crystals, covered on

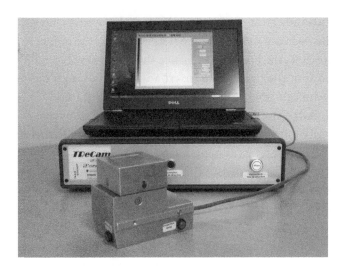

FIGURE 3.4   The intraoperative imaging probe tumour resection camera.

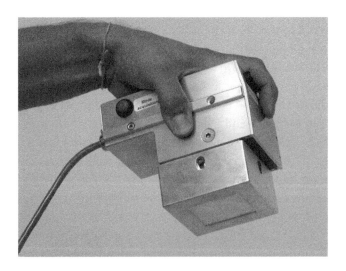

FIGURE 3.5   The detection head of tumour resection camera.

its top with Teflon optical coating and on its edges with absorbing surface. The optical coupling with MAPMT is ensured by a thixotropic gel (NyoGel OC-662, Smartgel).

The MAPMT is a Hamamatsu H9500 flat panel device. The input borosilicate glass window is 1 mm thick and the photocathode is a bialkali type (QE_{max} = 24% at 420 nm). The MAPMT is a 12-stage metal channel dynode (typical gain of 10^6) and 256 anode pixels. The effective area is 49 mm × 49 mm^2.

3.4.3 Electronic Read-Out

TReCam read-out includes four HAdronic Rpc Detector ReadOut Chip 2 (HARDROC2)-integrated devices that were designed in collaboration with the OMEGA/LAL.* The HARDROC2 read-out is a semi-digital read-out. The attributes of each of the 64 channels of HARDROC2, which are used in TReCam, are reported here:

1. A shaping unit based on a fast low-impedance preamplifier (adjustable gain over 8 bits per channel) followed by a slow shaper (peaking time, 50–150 ns) and a track and hold to provide a multiplexed analogue charge output up to 15pC

2. A trigger unit based on three adjustable gain fast shapers and three low offset discriminators (ranging from 10fC up to 10pC)

Each chip provides an analogue pulse burst, corresponding to the multiplexed output for the 64 individual charge measurements. They are coded by a 12 bit serial-to-parallel analog to digital converter (ADC). Data from the four ADCs are transferred to a field-programmable gate array (FPGA) for processing. Encoded data are then transmitted to a computer via USB link.

3.4.4 Image Processing

Transferred data, including image reconstruction and energy computations, were processed and displayed in real time by dedicated software developed in-house in C++ using QT.† Two algorithms are currently available to create the nuclear image: the 'raised to the power two', also called 'squared' algorithm introduced by Pani, and a method based on the least-square fit of the light distribution by a Cauchy function solved using Levenberg–Marquardt algorithm (Levenberg 1944; Marquardt 1963; Pani et al. 2009).

In order to reduce dead-time loss, the squared algorithm has been implemented directly on the FPGA. In this mode, only the gamma interaction position and the global energy of each event are transferred to the computer leading to a 700 cps maximum count rate before 20% loss of

---

* Pôle OMEGA du Laboratoire de l'Accélérateur Linéaire, Orsay, France.
† Released under Lesser General Public License.

counts, an improvement over the previous method which gave a lower maximum count rate of 170 cps.

TReCam exhibits the following performances at 140 keV: a spatial resolution of 1.8 mm, a sensitivity similar to POCI's of 300 cps/MBq and a much better energy resolution of 11%. TReCam is able to image the injection point and the SN without changing the camera settings.

3.4.5 Studies

We first designed an observational and feasibility study. We planned to enrol 30 breast cancer patients with non-palpable lesions, but only 20 were included between April 2011 and March 2014 in Lariboisière and Jean Verdier hospitals. The main goal was to check the concordance of the images obtained with the standard LS and TReCam. The secondary objectives were the evaluation of the radioimager in operating conditions (duration, image quality, acquisition difficulty) and criteria regarding the breast lesion: excised piece's volume, tumour margin quality, re-intervention rate and cosmetic results. The protocol was very similar to that of POCI's study except for the ^{99m}Tc-nanocolloid injection. A single ultrasound-guided radiologist performed all the injections, that is, two injections of 30 MBq each which is less than the half of the injected activity in POCI's study. The first injection was made at the superficial edge of the lesion and the second at the deeper edge of the lesion. Only 14 patients were eligible. All the results can be found in Bricou et al. (2015). As expected, LS, TReCam and counting probe showed 100% correlation finding the tumour (Figure 3.6).

The discrepancy is viewed in the SN localisation: the three devices agreed on location and number in 8/14 cases. For the 6 discordant cases, TReCam found more SNs than LS in 3/6 cases and LS more than TReCam in the remaining 3 cases. The lesion margins were *non in sano* or inadequate (less than 3 mm security margins) in 29% of the cases leading to 29% re-intervention rate. These four women all underwent a mastectomy due to multifocality or lesion size >4 cm. They should have been excluded from the trial, but due to the small size of the sample, we kept all the data. As in the POCI study, for one patient, the LS was completed with TReCam before the gamma probe–facilitated excision of an atypically located axillary SN. In that particular case, TReCam mapping of the dissection area probably saved time. Moreover, in one patient, LS failed to reveal the SN while TReCam detected two, which were confirmed by the gamma probe and removed.

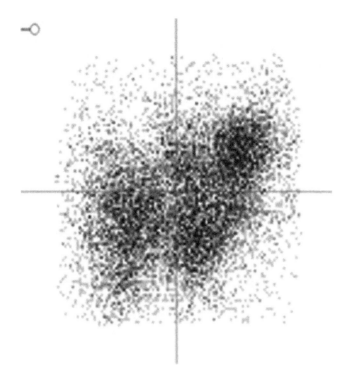

FIGURE 3.6　Tumour resection camera images the two injection sites around the tumour.

The impact of using TReCam during the SN procedure and tumour excision cannot be assessed by our small preliminary non-interventional study. That is why we have an ongoing trial where the end point is the rate of further surgery for inappropriate margins of the lumpectomy (less than 3 mm). This multicentric (three centres) phase II non-comparative interventional study will recruit 60 patients randomly distributed into 2 groups: one with a resection of the lesions according to the SNOLL procedure without the use of TReCam and the other with the use of TReCam.

3.5 CONCLUSION

POCI has shown its usefulness as a true substitute to the standard LS and a powerful tool in theatre for difficult SNB cases. TReCam is yet to prove its utility and efficacy in the SNOLL procedure. These handheld gamma cameras belong to a very large group of devices that are generating interest. Multimodality and 3D systems are developing rapidly and will soon make real-time gamma-imaging routine in the operating theatre.

REFERENCES

Aarsvold, J.N. and N.P. Alazraki. 2005. Update on detection of sentinel lymph nodes in patients with breast cancer. *Seminars in Nuclear Medicine*, 35(2):116–128.

Bernard, F., B. Geissler, F. Cachin et al. 2005. Captation des colloïdes par le ganglion sentinelle dans le cancer du sein: Comparaison entre deux techniques d'injection. *Médecine nucléaire*, 29(3):107–114.

Bricou, A., M.A. Duval, L. Bardet et al. 2015. Is there a role for a handheld gamma camera (TReCam) in the SNOLL breast cancer procedure? *Quarterly Journal of Nuclear Medicine and Molecular Imaging*. http://www. minervamedica.it/en/journals/nuclear-med-molecular-imaging/issue. php?cod=R39Y9999N00.

Bricou, A., M.A. Duval, Y. Charon and E. Barranger. 2013. Mobile gamma cameras in breast cancer care – A review. *European Journal of Surgical Oncology*, 39(5):409–416.

Callier, S., F. Dulucq, C. de La Taille et al. October 2007. Hardroc1, readout chip of the digital hadronic calorimeter of ILC. *IEEE Nuclear Science Symposium Conference Record*, Honolulu, Hawaii. Vol. 3, pp. 1851–1856.

Cousins, A., S.K. Thompson, A.B. Wedding and B. Thierry. 2014. Clinical relevance of novel imaging technologies for sentinel lymph node identification and staging. *Biotechnology Advances*, 32(2):269–279.

Feggi, L., E. Basaglia, S. Corcione et al. 2001. An original approach in the diagnosis of early breast cancer: Use of the same radiopharmaceutical for both non-palpable lesions and sentinel node localisation. *European Journal of Nuclear Medicine and Molecular Imaging*, 28(11):1589–1596.

Giacalone, P.L., A. Bourdon, P.D. Trinh et al. 2012. Radioguided occult lesion localization plus sentinel node biopsy (SNOLL) versus wire-guided localization plus sentinel node detection: A case control study of 129 unifocal pure invasive non-palpable breast cancers. *European Journal of Surgical Oncology*, 38(3):222–229.

Kawase, K., I.W. Gayed, K.K. Hunt et al. 2006. Use of lymphoscintigraphy defines lymphatic drainage patterns before sentinel lymph node biopsy for breast cancer. *Journal of the American Chemical Society*, 203(1):64–72.

Kerrou, K., S. Pitre, C. Coutant et al. 2011. The usefulness of a preoperative compact imager, a hand-held γ-camera for breast cancer sentinel node biopsy: Final results of a prospective double-blind, clinical study. *Journal of Nuclear Medicine*, 52(9):1346–1353.

Levenberg, K. 1944. A method for the solution of certain problems in least squares. *Quarterly of Applied Mathematics*, 1944(2):164–168.

Lovrics, P.J., S.D. Cornacchi, R. Vora, C.H. Goldsmith and K. Kahnamoui. 2011. Systematic review of radioguided surgery for non-palpable breast cancer. *European Journal of Surgical Oncology*, 37(5):388–397.

Marquardt, D.W. 1963. An algorithm for least-squares estimation of nonlinear parameters. *Journal of the Society for Industrial and Applied Mathematics*, 11(2):431–441.

Mathelin, C., S. Salvador, V. Bekaert et al. 2008. A new intraoperative gamma camera for the sentinel lymph node procedure in breast cancer. *Anticancer Research*, 28(5B):2859–2864.

Motomura, K., A. Noguchi, T. Hashizume et al. 2005. Usefulness of a solid-state gamma camera for sentinel node identification in patients with breast cancer. *Journal of Surgical Oncology*, 89(1):12–17.

Pani, R., F. Vittorini, M.N. Cinti et al. 2009. Revisited position arithmetics for LaBr 3: Ce continuous crystals. *Nuclear Physics B – Proceedings Supplements*, 197(1):383–386.

Pitre, S., L. Ménard, M. Ricard, M. Solal, J.R. Garbay and Y. Charon. 2003. A hand-held imaging probe for radio-guided surgery: Physical performance and preliminary clinical experience. *European Journal of Nuclear Medicine and Molecular Imaging*, 30(3):339–343.

Schwartz, G.F., A.E. Giuliano and U. Veronesi. 2002. Proceedings of the consensus conference on the role of sentinel lymph node biopsy in carcinoma of the breast, April 19–22, 2001, Philadelphia, Pennsylvania. *Cancer*, 94(10):2542–2551.

Scopinaro, F., A. Tofani, G. di Santo et al. 2008. High-resolution, hand-held camera for sentinel-node detection. *Cancer Biotherapy and Radiopharmaceuticals*, 23(1):43–52.

Thind, C.R., S. Tan, S. Desmond et al. 2011. SNOLL: Sentinel node and occult (impalpable) lesion localization in breast cancer. *Clinical Radiology*, 66(9):833–839.

Wendler, T., K. Herrmann, A. Schnelzer et al. 2010. First demonstration of 3-D lymphatic mapping in breast cancer using freehand SPECT. *European Journal of Nuclear Medicine and Molecular Imaging*, 37(8):1452–1461.

CHAPTER 4

Intraoperative 3D Nuclear Imaging and Its Hybrid Extensions

T. Wendler

CONTENTS

4.1 INTRODUCTION

Over the last 20 years, 3D nuclear medicine imaging, namely, single-photon emission computed tomography (SPECT) and positron emission computed tomography (PET), have become standard diagnostic imaging modalities in most institutions. In particular in oncology, nuclear medicine imaging (NM) allows the determination of targets for surgery. This is especially interesting if combined with anatomical imaging ('hybrid imaging') as in the SPECT/x-ray computed tomography (CT), PET/CT or lately PET/magnetic resonance imaging (MR). In NM/CT or NM/MR, the functional information can be put in the context of anatomy and thus aids in surgical planning and the definition of resection borders (e.g. in head–neck cancer [Nahas et al. 2005], in melanoma [Kumar et al. 2005], in ovarian cancer [Lenhard et al. 2008], among many other indications).

Nevertheless, it is still a major challenge to translate preoperative plans into the operating room. Today the imaging data mentioned earlier are commonly available in the operating theatre through the picture archiving and communication system – a system that is now a standard worldwide (Figure 4.1). However, during the intervention itself, these data can lose their validity to a large extent, due to changes in anatomy that start with different patient positioning, and extend up to changes due to resection and mobilisation of organs (e.g. in the abdomen [Baumhauer et al. 2008], in the brain [Hastreiter et al. 2004, Shamir et al. 2009] and in the liver [Maier-Hein et al. 2008]).

In order to account for this, in orthopaedics, ear–nose–throat surgery and neurosurgery, the so-called navigation systems, have been proposed (Peters 2006). These systems are based on the positioning systems that allow determination of the position and orientation of surgical instruments in relation to preoperative imaging data. In this way the relative position of the instruments can be displayed, and the surgeon can use this for 'navigating' them to the desired position in the anatomy. The preoperative data are thus brought directly to the operating site.

The use of these systems in combination with nuclear medicine imaging has been reported in neurosurgery (Sobottka et al. 2002), head and neck cancer (Feichtinger et al. 2010) and urology (Brouwer et al. 2014). However, the challenge still remains in how to compensate for deformation and changes in anatomy due to preparation and resection of structures.

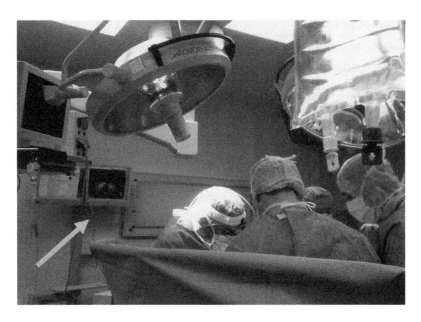

FIGURE 4.1   Preoperative positron emission computed tomography/x-ray computed tomography available over the picture archiving and communication system in the operating room during prostatectomy (arrow). Surgeons can always access the preoperative data; however, it is displayed far from the operating scene and it serves mostly as a static map.

An approach to account for changes before and during surgery is the use of intraoperative imaging. Intraoperative imaging can, in principle, exploit all the advantages of preoperative imaging and further allow its use during the surgical procedure in order to update plans and verify results after each single step, if necessary.

In the case of NM, intraoperative 3D imaging presents a number of constraints that makes it hard to simply place the diagnostic imagers to the operating room. The sole example of such an approach is the advanced multimodal image-guided operating suite which integrates in a single operating room a PET/CT, an MR and an angiography suite (Tempany et al. 2015). The major disadvantage of such a solution is the cost (several million dollars), the complicated logistics and the amount of space required.

A different approach was introduced in 2007 to enable 3D intraoperative nuclear imaging with flexible detectors (Wendler et al. 2007). In the following sections, this approach and its applications and extensions will be discussed in detail.

4.2 HISTORY

In 2005, our group at the Technische Universität München in Munich, Germany, came up with the idea of combining a handheld nuclear detector and position tracking in order to generate 3D nuclear images. After analysing the way SPECT images were generated, we realised that the limiting factor towards using SPECT in the operating theatre was the gantry. The gantry in SPECT devices holds the nuclear detectors and rotates them around the patient, while they obtain projection images needed for 3D image reconstruction.

By replacing this gantry with a position tracking system and using smaller nuclear detectors that could be handheld, a set of projection images could be acquired and possibly 3D images could be reconstructed.

The concept was not new; indeed, since 1989, scientists had already proposed using nuclear detectors and navigation systems (Hartsough et al. 1989). It was probably, however, Irving Weinberg who first drafted the concept of imaging using tracked radiation detectors (Weinberg 2003), with the first implementation by the group of José María Benlloch shortly after (Benlloch et al. 2004). Benlloch's concept was directly derived from the field of computer vision. His team had an operational mini-gamma camera, to which he added tracking markers in order to position it at particular orientations from an object and acquire projection images. The images generated by the camera in each position and the information on the position itself were then used to find correspondences between hotspots and calculate the 3D position of these by triangulation. The results ended up in two patent applications and a paper, but the project was abandoned.

Our idea, however, was slightly different from the implementation of Benlloch. We considered using the readings of a tracked gamma probe moved by hand, instead of gamma cameras, in order to generate an image. It took almost a year for us to get from the idea to a first prototype implementation, but by the beginning of 2007, the very first images were reported (Wendler et al. 2007). This new 'modality' was promptly baptised as freehand SPECT or handheld SPECT. Its approach can also be applied to PET, where one speaks of freehand PET. If both freehand SPECT and freehand PET are meant, then we use the name freehand NM.

4.3 BACKGROUND OF 3D IMAGE RECONSTRUCTION

To understand freehand NM, one needs to understand the process of generating fully tomographic data sets from projections. Tomography ('cross-sectional imaging') in medical imaging has been exploited widely in the past. CT, for example, uses a full set of 2D x-ray projections ('radiograms') acquired around the object of interest. This set of projections is then 'back projected' onto 3D volume in order to generate 3D images.

A similar approach is used in SPECT imaging. There the projections are planar scintigraphic images ('scintigrams'), and commonly, the algorithms used are more sophisticated than the back-projection methods that are still implemented in older CT or SPECT devices. The need for more complex algorithms to generate 3D images from a set of scintigrams stems from the fact that the mathematical assumptions for the projections diverge too much from reality.

Algorithms meant to generate 3D images from projections are often referred to as 3D reconstruction algorithms. The problem to be solved by those algorithms in mathematical terms is the determination of a function f in 3D given a (mathematical) set of projections $\{g_t\}$ in 2D where f is normally a scalar function of space. Each element of $\{g_t\}$ is, on its own, a 2D image taken at a time t and at a relative position x_t and orientation d_t when the 2D image was acquired in relation to the coordinate system used for f. Typical functions of f can be the x-ray attenuation (for CT) or the radioactivity concentration (for SPECT or PET). The images g_t are directly related to f, since one can see them as projection images of it. Finally, reconstruction algorithms can be stated as problems where the equation

$$\{g_t\} = \{h(f, x_t, d_t)\} \tag{4.1}$$

has to be inverted. The mapping h describes how the desired function f influences each of the acquired images $\{g_t\}$, that is the way function f is projected. Given that the devices generating the images work with discrete representations and that the mapping h is often assumed to be linear with respect to f, the previous equation turns into

$$\{g_i\} = \{H(x_i, d_i) \times f\} = \{H_i \times f\} = \left\{ \sum_j H_{i,j} \times f_j \right\} \tag{4.2}$$

where each g_i is a vector containing all pixel values of image i. The vector f contains the voxels of the volume to be reconstructed and H_i is the (projection) matrix that maps f to g_i.

Note that this equation is actually a set of linear equations and inverting this equation is equivalent to solving this set of equations. This process is called image reconstruction and, in the case of NM, is commonly accomplished using iterative reconstruction algorithms (in contrast to analytic algorithms that are applicable in particular cases but deliver less satisfactory results in general).

Iterative reconstruction algorithms find a solution by starting from an initial solution candidate and calculating a new candidate based on the previous candidate iteration per iteration. In most cases, as the input data are real and noise is not too high, these algorithms converge to a solution candidate which is close to reality within reasonable time (i.e. seconds to minutes).

In particular, in nuclear medicine imaging, most iterative reconstruction algorithms are derived from the maximum likelihood estimation maximisation (ML-EM) algorithm, an algorithm based on the principles made popular by Dempster, Laird and Rubin in the 1970s (Dempster et al. 1977) and corrected later by Wu (1983). Given that the image acquisition process can be very well modelled as a Poisson process, the ML-EM algorithm can be explicitly formulated as

$$f_j^{n+1} = \frac{f_j^n}{\sum_l H_{l,j}} \times \left[\sum_i H_{i,j} \frac{g_i}{\sum_k H_{i,k} f_k^n} \right] \tag{4.3}$$

where f_j^n is the value of voxel j at iteration n.

The matrix H contains elements $H_{i,j}$. Each element $H_{i,j}$ describes the influence on the image g_i of a unitary value on voxel f_j. Matrix H is commonly known as the system matrix since it models the complete set-up.

4.3.1 From Conventional Gantry-Based Imaging to Freehand Imaging

Dropping the gantry in NM dramatically changes the border conditions of the imaging problem (Table 4.1).

Most conventional NM imagers either use a ring of detectors (as in PET) or acquire projections from 360° around the patient (as in SPECT) (Wernick and Aarsvold 2004) to get a complete set of projections (i.e. full angular coverage). If the gantry is to be replaced by a handheld movement, this becomes an impossible task: because, first, the user cannot go all around the patient and, second, if no gantry is used, it is practically

TABLE 4.1   Comparison of Freehand NM and Conventional (Gantry-Based) NM

Property	Gantry-Based Imaging	Freehand Imaging
Calculation of relative position of projections	Per construction	Using tracking
Angular coverage of projections	Full angle	Limited angle
Symmetry of projections	Per construction	Impossible
Applicable reconstruction algorithm	Analytic or iterative	Iterative
Pre-computation/measurement of system matrix for iterative reconstruction	Possible	Impossible
Weight of detector	~1000 kg	0.2–2 kg
Field of view of detector	~30 × 30 cm^2	~5 × 5 cm^2
Statistics of acquired data	High	Low
Distance between detector and anatomy	Far	Close

impossible for a user to make trajectories with the symmetry of the gantry systems.

The lack of symmetry means the reconstruction algorithms cannot be analysed. The lack of a complete angular coverage results in a limited-angle reconstruction problem where special compensation steps are needed to minimise artefacts (Barrett 1990, Jaffe 1990). Yet in those approaches the images are commonly acquired in equidistant steps, following a symmetric trajectory using specialised gantries. More interesting is the case of non-circular trajectories used in state-of-the-art SPECT devices. In those cases, for each step, the detector is placed as close to the object of interest as possible. The resulting trajectory is similar to an ellipse but might not necessarily be one (Todd-Pokropek 1983, Abe et al. 1991). Still a freehand trajectory is far from a non-circular semi-elliptical trajectory.

Second, the use of gantries brings with it a further advantage: the region to be reconstructed can be easily defined. For most preoperative systems, the region to be reconstructed is merely a cylinder with the radius of the accessible space of the gantry and a length which is selected by the operator of the device depending on the anatomy to be scanned. Some systems include a 'topographic' image (i.e. image for means of orientation in relation to the topography patient) to help on the latter selection. That image is a very fast 2D anatomical image (e.g. in CT or MR devices) used to select the region to be reconstructed before a time-consuming or radiation-involving 3D acquisition.

Third, in conventional imagers, given that the symmetry is always the same, commonly the system matrix H is pre-computed by either analytical means or simulation (Moehrs et al. 2008, Ortuno et al. 2010), or it is measured directly using a known point source which is placed in different positions within the field of view (Panin et al. 2006). Further, since the reconstruction problem represents an ill-posed problem, it is of fundamental importance that the information provided in the system matrix is sufficient for a valid reconstruction (e.g. on the effect of errors in modelling on reconstruction, see Qi and Huesman [2005]).

If the gantry is to be eliminated, there are constraints applicable to the detectors. If an operator is meant to hold them in his/her hand during surgery, weight must be minimised (<1 kg). As a result, the detectors have to be small which impacts on sensitivity (the larger the detector, the higher the sensitivity), collimation (the bigger the collimator, the higher the collimation) and field of view.

The fact that smaller detectors are used also brings a major difference between conventional NM and freehand SPECT. Conventional NM systems collect millions of events per acquisition; a common acquisition using a handheld device provides a few hundred counts per second, resulting in a few thousand events per acquisition. The amount of collected events has a direct impact on the variance of the acquired images and as such impacts the signal-to-noise ratio.

Smaller detectors do, however, have a positive side. They can be placed very close to the anatomic region of interest or even into the surgical opening making the distance between detector and target significantly smaller when compared to the big detectors which tend to be far away from the patient in typical conventional NM. This aspect is, in contrast to all previous differences, an advantage of freehand NM over conventional NM.

4.4 IMPLEMENTATION OF FREEHAND NUCLEAR MEDICINE IMAGING

Given the differences with conventional (gantry-based) NM listed earlier, the implementation of freehand NM requires particular system hardware and algorithms.

4.4.1 Hardware

The major hardware components are shown in Figure 4.2. In general, any tracking system can be used for determining the position and orientation of the detector.

Currently, optical (e.g. NDI, ART) and electromagnetic (e.g. ASCENSION, POLHEMUS) tracking technologies dominate the market

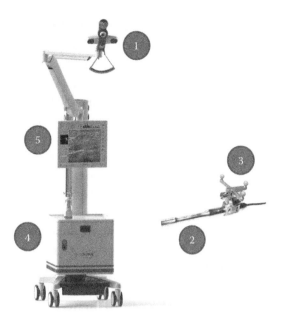

FIGURE 4.2 Hardware components needed for freehand single-photon emission computed tomography. A tracking system (here optical tracking system held by a flexible arm, 1) is used to determine the position and orientation of the nuclear detector (here a single-pixel gamma detector, also known as gamma probe, 2). In order to do so, a tracking target or sensor (here a passive tracking marker configuration, 3) is fixed to the gamma detector. All data are processed in a central processing unit (here a medical PC, 4) and displayed on a screen to the user (here medical touchscreen, for also user interaction, 5). In the implementation here, as offered by the company SurgicEye GmbH, the hardware is placed on a medical certified cart for flexible placement in the operating theatre.

of medical tracking systems; however, the increased use of robotic arms as mechanical trackers may change this situation (e.g. KUKA). These systems provide 3D position and 3D orientation information with submillimetre and sub-degree precision at sampling frequencies >20 Hz, which is sufficient for freehand NM. It is relevant to mention that it must be guaranteed that the tracking system does not interfere with the nuclear detector or vice versa.

On the detector side, there are currently available handheld detectors that are one-pixel gamma detectors (also known as gamma probes; Hoffman et al. 1999) or small gamma cameras (Vidal-Sicart et al. 2014). These detectors need to provide a stream of data with a frequency of >20 Hz to avoid motion blurring. Ideally also energy information of the events is useful in order to set energy windows to discard scattered events and sometimes Bremsstrahlung from true events.

In order to know where the detector is placed in space, the tracking system needs some way to identify the detector. To do this, a tracking target or sensor needs to be placed as close as possible to the detector, while avoiding handling/ergonomics issues. Placing a tracking target or a tracking sensor on a handheld detector does not completely fulfil the task of getting the detector readings in the same coordinate system as the image f. The process of calculating such transformations is commonly known as calibration (see Wendler et al. (2006) and Matthies et al. (2014) for examples on how to calibrate single-pixel and multiple-pixel detectors, respectively). If mechanical drawings for the tracking target or sensor mounted to the detector are available, the transformation mentioned earlier can be calculated and no calibration is needed.

Besides the tracking system (the detector and the tracking targets or sensors), a fast computer system is also needed. In freehand NM the requirement for fast image reconstruction displayed to the surgeon forces the use of high-end devices. It is important to mention that these computers need data acquisition interfaces for the tracking system and the nuclear detector.

Finally, a screen, or better a touchscreen, rounds off the hardware allowing the user to view the generated images and, in the case of a touchscreen, also provide intuitive user interaction.

4.4.2 Algorithms

The overall software architecture of freehand NM is depicted in Figure 4.3 illustrating the process of freehand SPECT with a one-pixel detector. For the following explanation and for the sake of simplicity, it is assumed that the volume of interest (i.e. the region to be reconstructed) is known.

First, the readings of the handheld detector g_i and its position x_s and orientation d_s need to be acquired and synchronised. Synchronisation minimises the dependence of the image quality on the motion of the detector. In order to synchronise these streams of data, common solutions are used to trigger signals by the hardware or timestamps and clock synchronisation are used (Wendler et al. 2006).

After synchronisation, the data are analysed and processed. The processing here is the merging of close readings (very similar positions and orientations) and discarding readings that do not contain information relevant to the volume of interest.

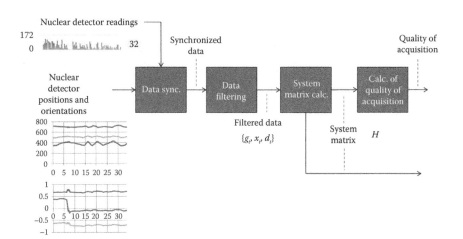

FIGURE 4.3   First steps of freehand single-photon emission computed tomography for a one-pixel detector. The nuclear detector feeds the system with a stream of reading, here counts per sampling time. The tracking system provides a stream of positions and orientations. The nuclear detector readings, positions and orientations are synchronised and filtered. Based on the filtered synchronised positions and orientations, the system matrix H is computed. The matrix H is also used to calculate the quality of the acquisition.

Since every acquisition is unique in freehand NM, so is its system matrix H. As stated earlier, each row H_i of the system matrix is calculated from the position x_i and orientation d_i of the detector at the moment image g_i is acquired. For this calculation, several options are available. The first is the use of analytical models for the response function of the detector (see Figure 4.4a and b) or Monte Carlo simulations (e.g. see Figure 4.4c and d). Another option is the measurement of the response using a point source and, for example, a 3D positioning table to generate a grid of readings (see Figure 4.4e and f).

It has been shown that it is useful to evaluate the quality of the acquisition (Wendler et al. 2010). We understand 'quality of the acquisition' here as a mask of the volume of interest showing which voxels of the volume of interest contain sufficient information for a valid reconstruction. More details on how to calculate the quality of an acquisition are given in Vogel et al. (2013). This mask is used to process the system matrix and later the reconstructed 3D image. In summary, voxels that do not have a minimum information content are not reconstructed (Figure 4.5).

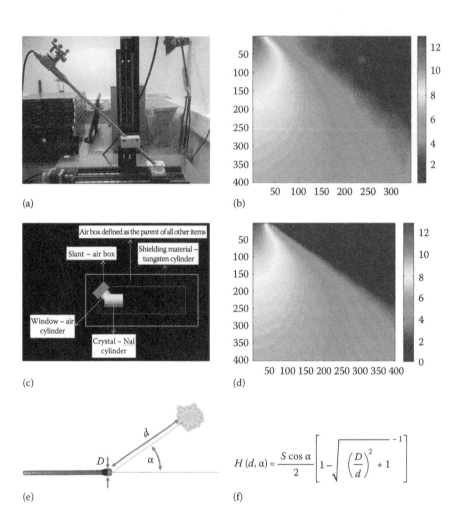

FIGURE 4.4 **(See colour insert.)** Different options for determining the response function of a nuclear detector (here a single-pixel detector). (a) Gamma probe fixed on a positioning table in order to move it relative to a point source and acquire a position by position response. (b) Response function acquired, for example, in (a), as a plane across the detector axis. (c) Simulation model for a gamma probe. (d) Response function simulated, for example, in (c), in a plane across detector. Here the point source was virtually moved in front of the detector. (e) Schematic for simple analytical model for response function. (f) Exemplary simple model based on (e) for response function of a single pixel cylindrical with diameter D and normalised sensitivity S.

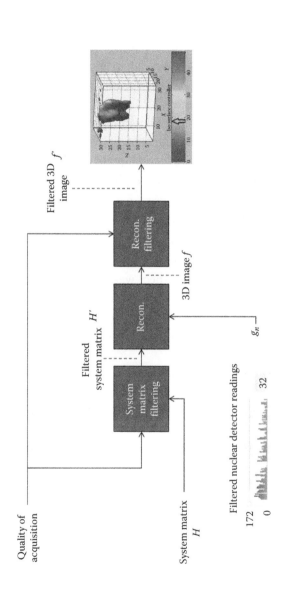

FIGURE 4.5   Final steps for freehand NM in the case of a single-pixel detector. The quality of the acquisition is used to discard voxels of the volume of interest ('system matrix filtering') before 3D reconstruction. For the reconstruction, the readings of the nuclear detector are needed. The resulting 3D image is further filtered also using the quality of the previously calculated acquisition.

In freehand NM, iterative reconstruction algorithms are used. These algorithms are applied to the processed system matrix H' and the filtered nuclear detector readings (all non-discarded detector readings) to generate an image f. Among the algorithms that have been shown to be useful in freehand NM, we have the classic ML-EM algorithm but also the one-step-late (Green 1990) or the separable paraboloidal surrogates (Erdogan and Fessler 1999) methods and neighbourhood-based Gibbs penalisation for regularisation. Depending on the algorithm used, the iteration number is commonly between 10 and 50.

Finally, the 3D reconstructed image is again processed. Commonly, an isotropic Gaussian filtering with a radius of 1–3 voxels produces satisfactory images. The mask with the acquisition quality is also used to recover the discarded voxels and enable proper visualisation.

4.4.3 Visualisation of Freehand NM Images

Freehand NM enables the generation of 3D NM images in the operating room; however, their value is minimal if these images are not put in the context of the operation sites. This aspect was clear from the very beginning for our group and already the first prototypes of freehand SPECT back in 2007 included an augmented reality visualisation where the freehand SPECT images where overlaid on the video of an optical camera situated over the operating sites.*

An important point to mention is that if a freehand NM image needs to be fused with another intraoperative imager (like a video camera or an ultrasound [US] device), it is recommendable to also track the patient. This can easily be achieved using a target or sensor taped onto the skin. An advantage of this is that the tracking target or sensor can be used as coordinate system for the freehand NM image and the image fusion can be implemented as the closing of a transformation chain.

Two options for visualisation of freehand NM images are described.

4.4.4 Freehand SPECT/Optical Combination

The first implementation of freehand SPECT included an optical camera for combined freehand SPECT/optical visualisation. The image fusion

---

* An augmented reality visualisation can be thought of as an image fusion of a non-optical 3D image and a 2D optical image. Here the non-optical 3D image is, of course, the freehand NM image.

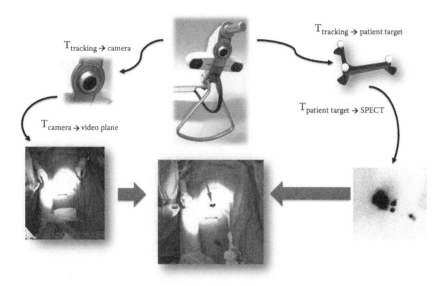

FIGURE 4.6   In the presented implementation, the tracking system is an optical tracking system. An optical camera is placed rigidly fixed to the tracking system. The calibration of the optical camera is done using the hand–eye calibration algorithm of Daniilidis. For freehand single-photon emission computed tomography (SPECT), if the image is reconstructed in the coordinate system of a patient target or sensor, then the tracking system provides the transformation from this target or sensor to the tracking system itself. The transformation from the patient target to the SPECT image can be predefined or defined at the beginning of the acquisition. In this way, the freehand SPECT can be translated and rotated to the coordinate system of the camera and finally projected on the image plane. The resulting image can be rendered such that the radioactive uptake above a certain threshold appears opaque on the video image and the uptake below is transparent as shown in the example.

requires essentially that both the freehand SPECT image and the optical video stream are on the same coordinate system. A possible implementation for this is shown in Figure 4.6.

In any augmented reality/optical image fusion, the optical camera needs to be calibrated. Calibration here has mainly to do with the definition of the projection matrix from the camera coordinate system to the image plane. A simple approach to achieve calibration is the calculation of the transformation from the camera coordinate system to the camera plane using known patterns as described in Hartley and Zisserman (2003).

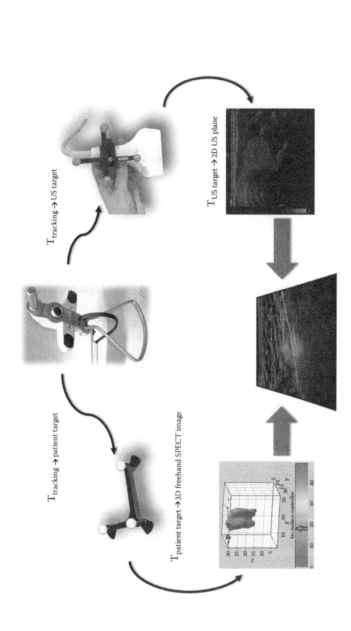

FIGURE 4.7 **(See colour insert.)** Image fusion of freehand single-photon emission computed tomography (SPECT) with ultrasound (US). An optical tracking system is used to track both the patient and the US probe using tracking targets. A previously acquired freehand SPECT image is already in the patient target coordinates. By translating and rotating it following the transformation chain shown here, the freehand SPECT image can be put in the coordinate system of the tracking target of the US. Then using the projection matrix that is obtained during the calibration, the freehand SPECT image can be overlaid on the US.

For defining the transformation from the camera to the tracking system, algorithms like the hand–eye calibration in Daniilidis (1999) can be used.

It is important to note that given the nature of freehand NM, it is a snapshot image; that is, the user needs to acquire data for reconstruction and wait for the system to complete the reconstruction as described earlier before being able to view it. If fused with an optical video stream, the video can be real time, but the freehand NM will not be.

In order to solve this latency, our group enables the visualisation of the reconstruction images while the reconstruction runs. A moving window reconstruction is implemented such that the user gets a 3D freehand NM image of the last 30–90 s every few seconds.

4.4.5 Freehand SPECT/US

A similar approach can be taken if the freehand NM images are meant to be fused with US. Here a tracking target or sensor can be placed on the US probe. In order to calibrate the US, several algorithms and phantoms are available. An example of these is the single-wall calibration algorithm as proposed by Prager et al. (1998).

As with the augmented reality visualisation, a cycle of transformation is needed as shown in Figure 4.7.

4.5 EXEMPLARY APPLICATIONS OF FREEHAND NM IMAGES

In the following section, I will describe exemplary clinical applications using freehand NM.

4.5.1 Freehand SPECT for Radioguided Resection

The first application of the freehand NM was the mapping of sentinel lymph nodes (SLNs). In these particular indications, lymph nodes showing an uptake of ^{99m}Tc-colloids are depicted. The contrast is commonly very high, as the ^{99m}Tc-colloid is only to be found in lymph nodes, lymphatic vessels and the injection site. Typical activities are between 20 kBq and 2 MBq in the lymph nodes, in a volume of roughly 0.5–3 cm^3 at depths between 1 and 6 cm. The activity in the injection site is commonly 20–300 MBq and its distribution is rather shallow and local to the injection points (normally 3–4 injections are used). On the anatomical side, due to the established indications for sentinel node mapping, lymph nodes are usually located in the axilla (breast and skin cancers), the groin (penis, vulva and skin cancers), the neck (head/neck and skin cancers) or deep in the pelvis or along the aorta (prostate, cervix and endometrial cancer).

Having 3D imaging in these indications has several advantages over the conventional technique of mapping the said nodes using acoustics:

1. The availability of intraoperative SLN mapping would enable the injection of the radioactive tracer intraoperatively, simplifying the logistics and making it possible to reduce the injected dose as preoperative imaging could be skipped – this would be of particular interest for indications where the injection is cumbersome, like in cervix or endometrium and deep head/neck cancers.

2. Having in situ visualisation and depth information aids planning the incision and guiding the surgeon to the sentinel nodes, minimising damage to the surrounding tissue – in particular this is valid in the neck and in pelvic/para-aortic lymph nodes.

3. Being able to repeat the SLN mapping at any time during surgery enables the control of the resection and the reassessment of plans on the fly in difficult cases, where the original plan presents problems – this can be useful in skin cancer and pelvic tumours.

4. Finally, in cases where the injection site is close to the SLNs, intraoperative 3D lymphatic mapping can help separate the injection site and nodes far more easily than with acoustics – this is applicable in vulva and head/neck cancer particularly.

The preferred implementation for freehand SPECT in this context is the use of a gamma probe as nuclear detector and an optical camera above the operation sites or laparoscope for image fusion. Figure 4.8 shows an example implementation for laparoscopic surgery. It has to be noted that if a laparoscope is used as the camera, it needs to be tracked on its own, which is different to that shown in Figure 4.6.

To date, such architecture for freehand SPECT/optical combination is in use in around 50 centres worldwide. Next to SLN mapping (in breast cancer (Wendler et al. 2010, Bluemel et al. 2013), head/neck cancer (Bluemel et al. 2014, Heuveling et al. 2015), melanoma (Mihaljevic et al. 2014)), applications also include radioguided occult lesion localisation (Bluemel et al. 2015), radioactive seed localisation (Pouw et al. 2014a,b), radioguided parathyroidectomy (Rahbar et al. 2012) and radioguided resection of paragangliomas (Einspieler et al. 2014), among others.

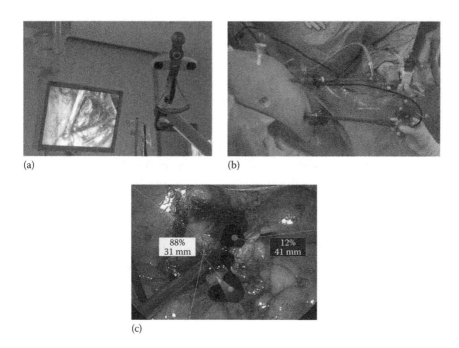

(a)

(b)

(c)

FIGURE 4.8 (a) Optical tracking placed over the operating sites (here next to the screen of the laparoscope) determines in real time the position and orientation of the instruments in (b): the laparoscopic gamma probe (nuclear detector, lower instrument) and the laparoscope (optical camera, higher instrument). (c) User can see in one screen the radioactive lymph nodes and the anatomy. Here two lymph nodes next to arteria iliaca interna can be seen together with the metallic shaft of the laparoscopic gamma probe. The weak sentinel lymph node (with 12% of the total radioactivity of the area) is at 41 mm from the tip of the gamma probe. The hot one with 88% of the activity is at a distance of 31 mm.

4.5.2 Freehand SPECT/US for Non-Surgical Sentinel Lymph Node Biopsy

In 2007, our group implemented an initial version of a freehand SPECT/US system; however, it was only in 2013 that such a system became available as a certified system. The concept was to implement the fusion depicted in Figure 4.7. The clinical motivation was to be able to see in real time on an US image which lymph node was a SLN. This visualisation would enable a needle biopsy of the node and potentially get the same diagnostic information as obtained in a surgical SLN resection.

Replacing the surgical biopsy of sentinel nodes with a needle procedure has several advantages:

1. A needle biopsy has a significantly lower complication rate than a surgical procedure since less damage is done to the adjacent tissue.

2. By converting the SLN biopsy into an ambulatory intervention, it can also be separated from the tumour resection; if this is the case, the information on the nodal status (target of the SLN biopsy) would be available before the tumour resection and be used for surgical planning.

3. Normally the SLN mapping is done the day before surgery so that the activity used has to be high enough to be detected the next day in the operating room. If, however, the SLN is sampled with a needle, the intervention can be started a few minutes after injection and/or mapping, and at least four times less radioactivity can be used.

The current implementation for freehand SPECT/US again used an optical tracking system and passive tracking targets on the gamma detector and the US probe. The gamma detector was a 4 cm² × 4 cm² handheld gamma camera with an 11 mm parallel-hole collimator. This type of detector has an advantage over gamma probes, having a higher sensitivity and enabling higher-quality images in less time. The needle is not tracked as it can be easily seen in the real-time US; a 10–13 G aspiration or freezing needles were used that enabled retrieval of up to 80% of the SLN in a few steps (Figure 4.9).

The initial results using freehand SPECT/US for SLN aspiration biopsy are encouraging (Freesmeyer et al. 2014a, Okur et al. 2014). In a set of yet unpublished results, patients of the ongoing MinimalSNB study (38 breast cancer patients, in 6 centres) had a needle biopsy taken from their SLNs under the guidance of freehand SPECT/US. All patients were indicated for a surgical biopsy which was performed immediately after the needle biopsy. Histopathological examination (H&E, step sectioning) of needle biopsies and surgically removed SLNs was compared. Final pathological examination of material harvested with both methods matched in 34 cases (33 negatives, 1 positive). The needle biopsy failed to detect metastases in two pN1 SLNs. In one case, the surgically resected tissue did not contain lymph nodes and the needle biopsy remained the only information on nodal status. In another case, a metastasis found in needle biopsy motivated a second reading of an originally negative

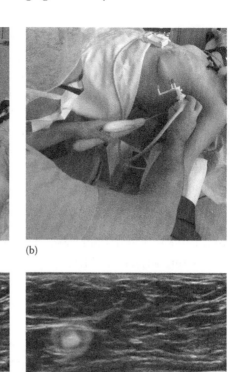

(a)

(b)

(c)

(d)

FIGURE 4.9 **(See colour insert.)** Freehand single-photon emission computed tomography (SPECT)/ultrasound (US) in action during sentinel lymph node (SLN) aspiration biopsy. (a) Freehand SPECT acquisition using a handheld gamma camera as nuclear detector, here in a breast cancer patient. (b) Placement of the needle for aspiration biopsy based on freehand SPECT/US images of (d). (c) B-mode image of axilla of patient showing at least one lymph node. (d) Freehand SPECT/US combination highlights the radioactive SLN by making it more prominent.

SLN which resulted in the upstaging of the patient. In both cases where metastases were missed by needle biopsy, the retrieved lymph tissue was minimal (1 × 14 G sample and 1 × 10 G sample tangential to node). Overall the SLN aspiration biopsy technique showed to be a valid method for percutaneous detection of SLNs and needle guidance.

Freehand SPECT/US is not only applicable to SLN aspiration biopsies; initial work has been reported also in thyroid (Freesmeyer et al. 2014b) and parathyroid imaging (Bluemel et al. 2015).

4.5.3 Freehand PET/US for Endoscopic Procedures

To close this chapter, a few paragraphs should be spent on freehand PET. In the initial patents of Weinberg in 2003, he postulated that both PET and SPECT could be acquired without a gantry with movable detectors (Weinberg 2003). In particular, for PET, the concept that was later adopted by several groups was to use a small detector inside the body in coincidence with a bigger detector or a ring outside the patient (Huh et al. 2007). It was, however, only in 2010 when the European Commission funded the Endoscopic Time-of-Flight PET/Ultrasound (EndoTOFPET/US) project that these ideas became a reality. The official acceptance came roughly 2 years after a meeting close to Mattinata in Italy, where Paul Lecoq and John Prior first asked our group to join them and develop our freehand SPECT into a freehand PET.

In the EndoTOFPET/US project, a small pixelated detector ('probe') is fixed to an endorectal US device, such that it can be placed inside the

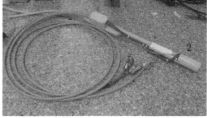

FIGURE 4.10 Images of Endoscopic Time-of-Flight PET/Ultrasound prototype after 5 years of work. The external plate (1) is a 23 × 23 cm² LYSO crystal array of 3.5 × 3.5 × 15 mm³ crystals and the probe (2) is a 13.5 × 13.5 mm² LYSO crystal array of 0.71 × 0.71 × 15 mm³ crystals. Both plate and probe use silicon multiplier sensors for detection. A group of three computers (3) take care of controlling the system (data acquisition, coincidence sorting, tracking system control, synchronisation, temperature control, etc.) and producing 3D images which are fused with the live stream from the ultrasound.

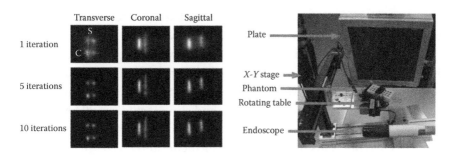

FIGURE 4.11   First images of Endoscopic Time-of-Flight PET/Ultrasound prototype acquired by the end of 2014 after 7 years of brainstorming and research and development. On the right, the experimental set-up is seen: two radioactive sources are placed between probe (bottom part of image) and plate (top part of image). The optical tracking system is not shown here. On the left, the reconstructed images are shown in the canonical planes for different iterations of a maximum likelihood estimation maximisation–adapted algorithm to be able to use time-of-flight information.

patient. The sensitive side of the bigger detector plate is placed on the opposite side of the patient such that the area to be imaged is located between probe and plate. Both probe and plate are tracked using an optical tracking system and calibrated such that the relative position and orientation of them are known. Using fast electronic read-outs and a coincidence module capable of time of flight, coincidences of probe and plate and the current relative position and orientation are acquired (see Figure 4.10). Following the concepts presented here, this data stream is then reconstructed using iterative algorithms to yield 3D images – for further details see Aubry et al. (2013).

To date, the EndoTOFPET/US prototype is under evaluation and still requires a significant amount of work before a prototype that can be used in humans is created. Nevertheless, the prospects are encouraging (Figure 4.11).

4.6 CONCLUSIONS

Intraoperative 3D NM is possible in a simple, flexible and cost-effective way using freehand NM. Freehand NM is a novel NM modality that generates NM-like images without the need of a gantry using handheld detectors. Given the need to put the freehand NM images in the context of surgery (or intervention), freehand NM is always combined with an additional imaging modality, to date, either optical cameras or US. Within this work, a brief overview was given on the background, characteristics, implementation and applications of freehand NM.

In particular, freehand SPECT has been shown to have the potential to bring relevant benefits to certain clinical indications. However, despite a few dozen publications, freehand SPECT and its hybrid extensions is still in its infancy. Technical improvements and clinical studies will be needed to make this technique a standard of care. The cornerstone is however laid and the current trends of personalised medicine and image-guided treatments will certainly push these developments further.

ACKNOWLEDGEMENTS

The work presented here is a summary of almost 10 years of research and development both at the Technische Universität München and the spin-off company SurgicEye GmbH in close collaboration with the Leiden University Medical Centre, the Universitätsklinikum Würzburg and the company Crystal Photonics GmbH, among many others. Further, the development of freehand PET was only possible due to the work of the EndoTOFPET/US collaboration. Over these years, a significant part of the funding came from European (EU FP7-HEALTH-2010-two-stage 256984, EU FP7 People 289355, Eurostars E!7103, Eurostars E!7555), German (BMBF MoBiTech 13N11143, ZIM EP130916) and Bavarian (BayTOU IBS-3667a/341/5-TOU-0902-0002, BayMed MED-1208-0003, BFS AZ-1072-13) public institutions.

Beyond the funding, this research would never have been possible without roughly 100 students (undergraduates, Masters, PhDs) and seasoned researchers. Among the names that need to be mentioned here, we have Prof. Nassir Navab, Prof. Sibylle Ziegler, Prof. Markus Schwaiger, Dr. Tobias Lasser, Dr. Joerg Traub, Stefan Wiesner, John-Michael Fischer and Àron Cserkaszky.

REFERENCES

Abe, S., M. Meguro and Y. Takeishi. September 1991. Evaluation of non-circular orbit in thallium-201 myocardial SPECT. *Kaku Igaku*, 28(9):1105–1109.

Aubry, N., E. Auffray, F.B. Mimoun et al. 2013. EndoTOFPET-US: A novel multimodal tool for endoscopy and positron emission tomography. *Journal of Instrumentation*, 8(04):C04002.

Barrett, H.H. 1990. Limited-angle tomography for the nineties. *Journal of Nuclear Medicine*, 31(10):1688–1692.

Baumhauer, M., M. Feuerstein, H.P. Meinzer and J. Rassweiler. 2008. Navigation in endoscopic soft tissue surgery: Perspectives and limitations. *Journal of Endourology*, 22(4):751–766.

Benlloch, J.M., M. Alcañiz, B. Escat et al. 2004. The gamma functional navigator. *IEEE Transactions on Nuclear Science*, 51(3):682–689.

Bluemel, C., A. Cramer, C. Grossmann et al. 2015. iROLL: Does 3-D radioguided occult lesion localization improve surgical management in early-stage breast cancer? *European Journal of Nuclear Medicine and Molecular Imaging*, 42(11):1692–1699.

Bluemel, C., K. Herrmann, U. Müller–Richter et al. 2014. Freehand SPECT-guided sentinel lymph node biopsy in early oral squamous cell carcinoma. *Head & Neck*, 36(11):E112–E116.

Bluemel, C., A. Schnelzer, A. Okur et al. 2013. Freehand SPECT for image-guided sentinel lymph node biopsy in breast cancer. *European Journal of Nuclear Medicine and Molecular Imaging*, 40(11):1656–1661.

Brouwer, O.R., N.S. van den Berg, H.M. Mathéron et al. 2014. Feasibility of intraoperative navigation to the sentinel node in the groin using preoperatively acquired single photon emission computerized tomography data: Transferring functional imaging to the operating room. *The Journal of Urology*, 192(6):1810–1816.

Daniilidis, K. 1999. Hand-eye calibration using dual quaternions. *International Journal of Research and Review*, 18(3):286–298.

Dempster, A.P., N.M. Laird and D.B. Rubin. 1977. Maximum likelihood from incomplete data via the EM algorithm. *Journal of the Royal Statistical Society: Series B (Methodological)* 39:1–38.

Einspieler, I., A. Novotny, A. Okur, M. Essler and M.E. Martignoni. 2014. First experience with image-guided resection of paraganglioma. *Clinical Nuclear Medicine*, 39(8):e379–e381.

Erdogan, H. and J.A. Fessler. 1999. Ordered subsets algorithms for transmission tomography. *Physics in Medicine and Biology*, 44(11):2835.

Feichtinger, M., M. Pau, W. Zemann, R.M. Aigner and H. Kärcher. 2010. Intraoperative control of resection margins in advanced head and neck cancer using a 3D-navigation system based on PET/CT image fusion. *Journal of Cranio-Maxillo-Facial Surgery*, 38(8):589–594.

Freesmeyer, M., T. Opfermann and T. Winkens. 2014b. Hybrid integration of real-time US and freehand SPECT: Proof of concept in patients with thyroid diseases. *Radiology*, 271(3):856–861.

Freesmeyer, M., T. Winkens, T. Opfermann, P. Elsner, I. Runnebaum and A. Darr. 2014a. Real-time ultrasound and freehand-SPECT. *Nuklearmedizin*, 53(6):259–264.

Green, P.J. 1990. Bayesian reconstructions from emission tomography data using a modified EM algorithm. *IEEE Transactions on Medical Imaging*, 9(1):84–93.

Hartley, R. and A. Zisserman. 2003. *Multiple View Geometry in Computer Vision*, 2nd edn. Cambridge University Press, New York.

Hartsough, N.E., H.B. Barber, J.M. Woolfenden, H.H. Barrett, T.S. Hickernell and D.P. Kwo. August 1989. Probes containing gamma radiation detectors for in vivo tumor detection and imaging. In *OE/LASE'89*, Los Angeles, CA. 15–20 January 1989, pp. 92–99.

Hastreiter, P., C. Rezk-Salama, G. Soza et al. 2004. Strategies for brain shift evaluation. *Medical Image Analysis*, 8(4):447–464.

Heuveling, D.A., S. van Weert, K.H. Karagozoglu and R. de Bree. 2015. Evaluation of the use of freehand SPECT for sentinel node biopsy in early stage oral carcinoma. *Oral Oncology*, 51(3):287–290.

Hoffman, E.J., M.P. Tornai, M. Janecek, B.E. Patt and J.S. Iwanczyk. 1999. Intraoperative probes and imaging probes. *European Journal of Nuclear Medicine and Molecular Imaging*, 26(8):913–935.

Huh, S.S., N.H. Clinthorne and W.L. Rogers. 2007. Investigation of an internal PET probe for prostate imaging. *Nuclear Instruments and Methods in Physics Research Section A: Accelerators, Spectrometers, Detectors and Associated Equipment*, 579(1):339–343.

Jaffe, J.S. 1990. Limited angle reconstruction using stabilized algorithms. *IEEE Transactions on Medical Imaging*, 9(3):338–344.

Kumar, R., A. Mavi, G. Bural and A. Alavi. 2005. Fluorodeoxyglucose-PET in the management of malignant melanoma. *Radiologic Clinics of North America*, 43(1):23–33.

Lenhard, M.S., A. Burges, T.R. Johnson et al. 2008. PET-CT in recurrent ovarian cancer: Impact on treatment planning. *Anticancer Research*, 28(4C):2303–2308.

Maier-Hein, L., A. Tekbas, A. Seitel et al. 2008. In vivo accuracy assessment of a needle-based navigation system for CT-guided radiofrequency ablation of the liver. *Medical Physics*, 35(12):5385–5396.

Matthies, P., J. Gardiazabal, A. Okur, J. Vogel, T. Lasser and N. Navab. 2014. Mini gamma cameras for intra-operative nuclear tomographic reconstruction. *Medical Image Analysis*, 18(8)1329–1336.

Mihaljevic, A.L., A. Rieger, B. Belloni et al. 2014. Transferring innovative freehand SPECT to the operating room: First experiences with sentinel lymph node biopsy in malignant melanoma. *European Journal of Surgical Oncology*, 40(1):42–48.

Moehrs, S., M. Defrise, N. Belcari et al. 2008. Multi-ray-based system matrix generation for 3D PET reconstruction. *Physics in Medicine and Biology*, 53(23):6925.

Nahas, Z., D. Goldenberg, C. Fakhry et al. 2005. The role of positron emission tomography/computed tomography in the management of recurrent papillary thyroid carcinoma. *The Laryngoscope*, 115(2):237–243.

Okur, A., C. Hennersperger, B. Runyan et al. 2014. FhSPECT-US guided needle biopsy of sentinel lymph nodes in the axilla: Is it feasible? *MICCAI 2014*, Boston, MA. pp. 577–584.

Ortuno, J.E., G. Kontaxakis, J.L. Rubio, P. Guerra and A. Santos. 2010. Efficient methodologies for system matrix modelling in iterative image reconstruction for rotating high-resolution PET. *Physics in Medicine and Biology*, 55(7):1833.

Panin, V.Y., F. Kehren, C. Michel and M. Casey. 2006. Fully 3-D PET reconstruction with system matrix derived from point source measurements. *IEEE Transactions on Medical Imaging*, 25(7):907–921.

Peters, T.M. 2006. Image-guidance for surgical procedures. *Physics in Medicine and Biology*, 51(14):R505.

Pouw, B., L.J. de Wit-van der Veen, D. Hellingman et al. 2014b. Feasibility of preoperative (125)I seed-guided tumoural tracer injection using freehand SPECT for sentinel lymph node mapping in non-palpable breast cancer. *European Journal of Nuclear Medicine and Molecular Imaging* 4:19.

Pouw, B., J.A. van der Hage, M.J.T.V. Peeters, J. Wesseling, M.P. Stokkel and R.A. Valdés Olmos. 2014a. Radio-guided seed localization for breast cancer excision: An ex-vivo specimen-based study to establish the accuracy of a freehand-SPECT device in predicting resection margins. *Nuclear Medicine Communications*, 35(9):961–966.

Prager, R.W., R.N. Rohling, A.H. Gee and L. Berman. 1998. Rapid calibration for 3-D freehand ultrasound. *Ultrasound in Medicine and Biology*, 24(6):855–869.

Qi, J. and R.H. Huesman. 2005. Effect of errors in the system matrix on maximum a posteriori image reconstruction. *Physics in Medicine and Biology*, 50(14):3297.

Rahbar, K., M. Colombo-Benkmann, C. Haane et al. 2012. Intraoperative 3-D mapping of parathyroid adenoma using freehand SPECT. *European Journal of Nuclear Medicine and Molecular Imaging*, 2(1):51.

Shamir, R.R., L. Joskowicz, S. Spektor and Y. Shoshan. 2009. Localization and registration accuracy in image guided neurosurgery: A clinical study. *International Journal of Computer Assisted Radiology and Surgery*, 4(1):45–52.

Sobottka, S.B., J. Bredow, B. Beuthien-Baumann, G. Reiss, G. Schackert and R. Steinmeier. 2002. Comparison of functional brain PET images and intraoperative brain-mapping data using image-guided surgery. *Computer Assisted Surgery*, 7(6):317–325.

Tempany, C., J. Jayender, T. Kapur et al. 2015. Multimodal imaging for improved diagnosis and treatment of cancers. *Cancer*, 121(6):817–827.

Todd-Pokropek, A. 1983. Non-circular orbits for the reduction of uniformity artefacts in SPECT. *Physics in Medicine and Biology*, 28(3):309.

Vidal-Sicart, S., M.E. Rioja, P. Paredes, M.R. Keshtgar and R.A. Valdés Olmos. 2014. Contribution of perioperative imaging to radioguided surgery. *Quarterly Journal of Nuclear Medicine and Molecular Imaging*, 58(2):140–160.

Vogel, J., T. Lasser, J. Gardiazabal and N. Navab. 2013. Trajectory optimization for intra-operative nuclear tomographic imaging. *Medical Image Analysis*, 17(7):723–731.

Weinberg, I.N., PEM Technologies, Inc. 2003. Hand held camera with tomographic capability. U.S. Patent 6,628,984.

Wendler, T., A. Hartl, T. Lasser et al. 2007. Towards intra-operative 3D nuclear imaging: Reconstruction of 3D radioactive distributions using tracked gamma probes. In *MICCAI 2007*, Brisbane, Australia, pp. 909–917.

Wendler, T., K. Herrmann, A. Schnelzer et al. 2010. First demonstration of 3-D lymphatic mapping in breast cancer using freehand SPECT. *European Journal of Nuclear Medicine and Molecular Imaging*, 37(8):1452–1461.

Wendler, T., J. Traub, S.I. Ziegler and N. Navab. 2006. Navigated three dimensional beta probe for optimal cancer resection. In *MICCAI 2006*, Copenhagen, Denmark, pp. 561–569.

Wernick, M.N. and J.N. Aarsvold. 2004. *Emission Tomography: The Fundamentals of PET and SPECT*. Academic Press, London, U.K.

Wu, C.J. 1983. On the convergence properties of the EM algorithm. *The Annals of Statistics* 11:95–103.

FURTHER READING

http://www.ar-tracking.com/technology/.
http://www.ascension-tech.com/products/.
http://www.kuka-healthcare.com/en/projects_studies/.
http://www.ndigital.com/products/#optical-measurement-systems.
http://polhemus.com/applications/electromagnetics.

CHAPTER 5

Clinical Requirements and Expectations

Using Sentinel Lymph Node Biopsy as an Example

C. Hope, R. Parks and K.-L. Cheung

CONTENTS

5.1 INTRODUCTION

Nuclear medicine technology has a number of different applications. This chapter discusses the use of gamma cameras in a clinical setting using the example of sentinel lymph node biopsy (SLNB), a diagnostic procedure now routinely used in breast cancer surgery. Breast cancer affects one in eight women in the United Kingdom and as such is highly topical. Despite the dramatic developments in the understanding and treatment options during the last four decades, there is still scope for improvement, particularly in the accurate assessment of type and scale of disease.

5.2 HISTORY OF SENTINEL LYMPH NODE BIOPSY

Variation in the role of SLNB is seen in practice across the globe with no international consensus for overall methodology. This section summarises the history of SLNB and the crucial stages in its development Tanis et al. 2001, D'Angelo-Donovan et al. 2012 (Figure 5.1).

The foundation of SLNB is the understanding of the lymphatic drainage of breast tissue. The function of the lymphatic system itself is a relatively

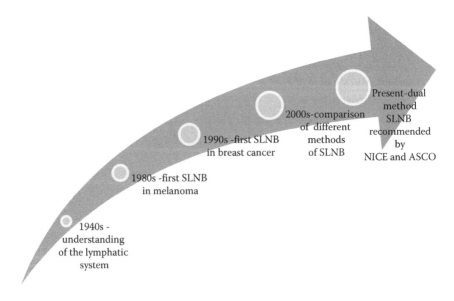

FIGURE 5.1   A time line of the history of sentinel node biopsy.

late discovery in the history of medicine. Today, it is well documented that the most prominent site of lymphatic drainage of the breast is to the axilla. A small percentage of patients will spread via alternate routes including the internal mammary, intramammary, supraclavicular, infraclavicular or contralateral axillary nodes.

We now know that tumour progression in breast cancer often occurs in an orderly, progressive fashion along the lymphatic system to the sentinel or gatekeeping node. The sentinel lymph node is known as the 'gatekeeper' node; it is the first node that receives metastatic deposits from the primary cancer site. So in theory, if the sentinel node is tumour-free, then the rest of the nodes following the lymphatic progression should also be tumour-free. SLNB involves identifying the sentinel node, removing the node and analysing it for malignant cells. The aim of this is to stage the axilla, to detect if the cancer in the breast has spread or if it has remained localised. The status of the axilla has been shown to be one of the most important prognostic indicators of the overall survival in breast cancer patients, and thus, the results of SLNB convey important information regarding further treatment options.

Before the advent of SLNB, removal of all of the axillary lymph nodes by axillary lymph node dissection (ALND) or examination of a set number of lymph nodes in the axilla by axillary node sampling (ANS) was the mainstay of axillary surgery for breast cancer treatment. There was no method of confirming the presence of axillary disease other than histological detection after surgical removal. Whereas ALND provided important prognostic indicators, it also carried a significant complication rate. With improved surgical techniques and greater patient expectation, a much-needed improvement in this area was required, particularly to reduce surgical morbidity.

5.2.1 1940s

In the 1940s, the understanding of the spread of cancer cells from the primary cancer site via the lymphatic system was discovered. Initial studies at that time focused on cancer of the parotid gland, penis and skin and the pattern of lymphatic drainage observed. The first description of a pathological sentinel lymph node following a total parotidectomy was made by Gould in 1951 (Gould et al. 1960). In the following two decades, lymphatic drainage from the penis and testicles was studied by Chiappa and Cabanas (Chiappa et al. 1966, Cabanas 1977).

5.2.2 1980s

Lymphatic mapping was becoming a commonly used method in the field of melanoma surgery in the late 1970s and the technique of intraoperative lymphatic mapping using blue dye was presented at the World Health Organisation's Second International Conference on Melanoma in 1989 (Morton et al. 1992, and Morton et al. 1989). Following on from melanoma, similar studies began to develop focusing on breast cancer. This led to studies of breast lymphatics by Haagensen and Christensen who identified the most common draining nodes of the breast (Haagensen 1972, Christensen et al. 1980).

5.2.3 1990s

Giuliano et al. were the first to publish the use of SLNB in the context of breast cancer surgery in 1994. Giuliano et al. performed 174 mapping procedures using a dye injected at the primary breast cancer site. The dye was taken up by the lymphatic system, and axillary lymphatics were identified and followed to the first sentinel node. Identification of the sentinel node accurately predicted axillary nodal status in 95.6% of cases (Giuliano et al. 1994). Following this, ongoing research regarding the use of SLNB over other forms of axillary surgery was explored.

5.2.4 2000s

During the 2000s, further evidence supporting the use of SLNB as opposed to a more drastic axillary surgery emerged (Glechner et al. 2013, Macaskill et al. 2012, Chetty et al. 2000). Veronesi et al. (2003) randomised 516 patients with primary breast cancer in whom the tumour was less than or equal to 2 cm in diameter to either SLNB ($n = 259$) or ALND ($n = 257$). The number of sentinel nodes detected was the same in the two groups; however, there was less pain and better arm mobility in patients who underwent SLNB alone. Among the patients who did not undergo ALND, there were no cases of overt axillary metastasis during follow-up. The investigators concluded that SLNB accurately predicted axillary pathology in 96.9% of patients with a false-negative rate of 8.8%.

Mansel et al. (2006) conducted a multicentre randomised trial in clinically node-negative breast cancer patients comparing those who underwent SLNB ($n = 515$) to those who received a standard axillary staging procedure ($n = 516$). SLNB was associated with reduced complications over a 12-month period, with no compromise in efficacy, measured by axillary recurrence rate, local recurrence and survival. The trial therefore recommended that the dual method of SLNB should be preferentially utilised in suitable patients. This led to the development of a formal training programme, the 'New Start', by the Royal College of Surgeons of England, which has been approved by the National Institute for Health and Care Excellence (NICE) guidance and is currently implemented in UK practice today Mansel et al. 2013.

5.2.5 Present

According to most recent NICE guidance in the United Kingdom (Yarnold 2009, clinical guideline 80) published in 2009 regarding patients with early invasive breast cancer with no evidence of cancer spread to the axilla, 'minimal surgery, rather than lymph node clearance, should be performed to stage the axilla'.

Across Europe, the European Association of Nuclear Medicine Oncology Committee with the Society of Nuclear Medicine and Molecular Imaging and the European Society of Surgical Oncology have endorsed SLNB as the staging procedure of choice, rather than ALND, in patients with early-stage biopsy-proven breast cancer without proven axillary lymph node metastases (Giammarile et al. 2013, Lyman et al. 2014).

American practice has concurred with that of the United Kingdom, with the American Society of Clinical Oncology recommending in 2014 that women without sentinel lymph node metastases should not routinely receive ALND.

In summary, current evidences from the United Kingdom, the rest of Europe and United States suggest that SLNB rather than ALND should be used in suitable candidates. Of course, further extensive treatment should be provided to those with preoperative ultrasound or histological evidence of metastatic disease or in those with positive sentinel lymph nodes; however, literature suggests that this does not necessarily mean ALND (Giuliano et al. 2011, Donker et al. 2014, Lyman et al. 2014).

5.3 CLINICAL REQUIREMENTS

Given the critical clinical significance of a positive SLNB, that is, negating the need for further axillary surgery, it becomes even more important to detect the sentinel lymph nodes by an accurate and precise measure. This leads to the clinical requirements for SLNB and those for its further development in order to ensure all patients are given similar quality of care regardless of where they are treated and set standards to minimise the risk of false-negative results and to limit unnecessary axillary surgery and prevent the subsequent complications.

5.3.1 Standard Approach

For the rest of this chapter, we will focus on SLNB as practiced in the United Kingdom due to the well-established breast screening programme and breast cancer–specific clinics across the country, alongside the background of its evolution as described earlier. The approach to SLNB is standardised across the United Kingdom, ensuring it is the same for all patients. This is crucial in ensuring that patients with breast cancer go on to receive the correct treatment following their initial staging procedure. It is performed across all breast units in the United Kingdom, and therefore, the procedure needs to be reproducible by different surgeons. A localisation rate of greater than 90% and a false-negative rate less than 10% have been set as the benchmark standards for SLNB by NICE. Some studies reported higher localisation rates of 98% in practice (Mansel et al. 2006).

5.3.2 Procedure

The procedure is summarised in Figure 5.2. The day before or on the same day of SLNB, a radioisotope is injected into the breast tissue. The most commonly used radioisotope is ^{99m}Tc. Preoperative lymphoscintigraphy used to be performed during the training phase (see New Start) but has since been stopped as a routine since surgeons are now very familiar with the procedure. During the operation, the blue dye is injected into the breast tissue around the nipple. The radioisotope and blue dye follow the lymphatic system to the sentinel node. The radioisotope and blue dye are taken up by the lymphatic tissues. A gamma probe is moved over the breast/axilla to detect radioactivity and its level is shown on a monitor. The surgeon moves the probe to the axilla and, using the highest readings of radioactivity, guides the incision. The axilla is opened and the tissue dissected; some visible blue nodes may be seen straight away and these are removed and sent for histological analysis. Sometimes, the blue nodes are not obviously visible, so the gamma probe is used to detect the point of highest radioactivity within the axilla and the surgeon uses the probe to target further dissection to find the nodes. The nodes with the highest radioactivity count are known as 'hot nodes' and

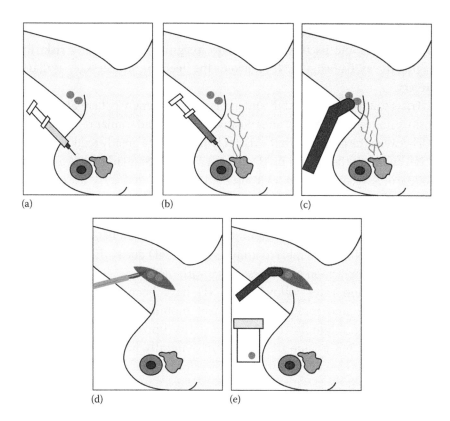

(a) (b) (c)

(d) (e)

FIGURE 5.2 **(See colour insert.)** The procedure of sentinel lymph node biopsy. (a) Radioisotope is injected into the periareolar region before the operation. (b) Blue dye is injected into the periareolar region on the day of SLNB and spreads via lymphatics to the axilla. (c) Gamma probe is used to detect radioactivity uptake in the axilla. (d) Using the gamma probe as a guide, an incison is made. Blue nodes can be seen. (e) The gamma probe helps to identify deeper sentinel nodes and these are removed for analysis.

are removed. Under normal circumstances, no more than three nodes should be removed during SLNB.

Depending on the histological results of the lymph nodes, further treatment is planned. If the nodes show no features of malignancy, no further treatment of the axilla is required. Histologically positive nodes, that is, showing signs of malignancy, require further treatment in the form of ALND, where all of the lymph nodes in the axilla are removed by surgery, though radiotherapy is another therapeutic option.

5.3.3 Decrease in Surgical Morbidity

Breast cancer and its treatment convey a significant morbidity risk. By staging the axilla using SLNB, it avoids the need for unnecessary axillary surgery.

Trials that compared SLNB and axillary radiotherapy (where SLNB had positive nodes), with ALND after positive SLNB, found similar regional control between the groups with the risk of lymphoedema greatly reduced in the SLNB and radiotherapy group (Donker et al. 2014). Similarly, the American College of Surgeons Oncology Group Z0011 trial found that the use of SLNB compared with ALND in patients with limited sentinel node positive breast cancer treated with breast conservation post-operative radiotherapy and systemic therapy did not result in an inferior survival (Giuliano et al. 2011). Therefore, SLNB is preferable to unnecessary ALND due to the higher morbidity and complication profile associated with the latter.

Complication rates following ALND have been quoted up to as much as 80% and include limited arm mobility, seroma formation, sensory changes, lymphoedema and negative impact on quality of life. Practically, SLNB as opposed to ALNB generally has a shorter hospital stay, quicker recovery time and, therefore, reduced overall cost per patient (Table 5.1).

5.3.4 Combined Technique

Sentinel node biopsy can be performed using radiopharmaceutical or blue dye administration or a combination of both techniques. The question then arises as to which of the three techniques provides the best result. McMasters et al. (2000) conducted the first multi-institutional study comparing blue dye versus radiopharmaceutical administration versus the two procedures combined. They demonstrated that combination of the two methods gave the highest percentage of localisation of the

TABLE 5.1   Common Complication Rates of Sentinel Lymph Node Biopsy versus Standard Axillary Treatment

	SLNB	Standard Axillary Treatment[a]
Lymphoedema (12 months post-op)	5%	13%
Sensory loss (12 months post-op)	11%	31%
Returned to work (1 week post-op)	31%	23%
Length of hospital stay (days)	4.1	5.4

Source: Mansel, R.E. et al., *J. Natl. Canc. Inst.*, 98(9), 599, 2006.

[a] Includes axillary lymph node dissection and axillary sampling.

sentinel lymph node and the lowest false-negative rate. This dual technique is recommended by NICE due to such increase in sensitivity.

Evidence to support this includes the National Surgical Adjuvant Breast and Bowel Project trial which found that SLNB by the combined method confirmed nodal status in 97.2% of participants with a false-negative rate of 9.8% (Krag et al. 2007). After 8 years of follow-up, this study showed that both disease-free survival and regional control of the disease were statistically similar in both groups.

5.3.5 Sensitivity and Specificity

NICE states that SLNB should be expected to have a >90% identification rate and a <10% false-negative rate. There is some room for improvement particularly in the false-negative rate. Patients that receive a false-negative result are understaged and therefore undertreated. This may result in the potential development of advanced metastatic and therefore incurable disease. Occasionally, SLNB localisation fails and an alternate procedure is then carried out. This is called axillary node sampling. ANS involves removing four axillary lymph nodes, normally from the low axilla, and sending them off for histological analysis. Factors that influence successful SLNB localisation include obesity, location of tumour and previous surgery or radiotherapy that may damage the lymphatic system as outlined in the ALMANAC validation phase study (Goyal et al. 2006).

5.4 EXPECTATIONS AND RECENT DEVELOPMENTS

Our understanding of breast cancer pathology is evolving, and subsequently, the expectations of the technology used to diagnose and treat breast cancer are also changing. When combined with evolving systemic therapies, treatment of axillary disease is now becoming less aggressive and with lower morbidity. Further developments in the field of SLNB include the optimisation of methods, timing of surgery and intraoperative examination of the histological specimens.

5.4.1 Optimisation of Equipment

Since the original SLNB technique was described, adaptations and improvements have been made. Initially, a cabled device attached to a monitor limited the surgeon's view of the radioactivity count and impacted upon the economy of movement. Recent advances have enabled the development of wireless handheld gamma probes that allow the surgeon a greater economy of movement. There are a wide variety of different shaped probes available to maximise operator view of the tissue and ease of localisation.

5.4.2 Intraoperative Examination of Sentinel Nodes

A further development that goes hand in hand with SLNB is the arrival of the RD-100i one-step nucleic acid amplification (OSNA) system and Metasin test (which targets breast epithelial cell markers cytokeratin-19 [CK19] and mammaglobin mRNA and identifies metastatic disease in sentinel lymph nodes). Both tests allow intraoperative detection of axillary metastases within the sentinel nodes. The advantage of intraoperative diagnosis is that patients can go on to have definitive surgical treatment immediately during the same anaesthetic, avoiding the need to wait for laboratory histological analysis. Further treatment such as chemotherapy and radiotherapy can also be initiated sooner. Intraoperatively, the taken sentinel lymph nodes are placed in the OSNA machine and the tissue is homogenised to allow analysis of the mRNA. The CK19 gene, an epithelial marker of breast cancer, is measured and the result based on the level of gene expression. This test takes approximately 45 min to perform. OSNA results in destruction of the node tissue so it cannot be analysed again. A systematic review concluded that OSNA was not cost-effective and was less accurate than histopathology in evaluating sentinel node metastases (Huxley et al. 2015). The importance of accurate detection of a positive sentinel lymph node and the development of improved cost-effective intraoperative imaging emerge.

The Metasin test uses a similar method but detects CK19 and mammaglobin by a quantitative reverse transcription polymerase chain reaction, both expressed in high levels by breast cancer. Some breast cancers do not produce the biological markers detected by the OSNA and Metasin tests; therefore, some people recommend preoperative tests of the diagnostic biopsies. NICE (diagnostic guidance DG8 2013) recommends OSNA as an option for the detection of intraoperative sentinel lymph node metastasis in those with early breast cancer and who would be a fit candidate for axillary node clearance. Currently, the Metasin test needs further more robust evidence before it can be incorporated into routine clinical practice.

5.4.3 Dual Technique

Currently, SLNB relies on a dual technique in order to achieve a satisfactory level of sensitivity and specificity as outlined earlier. This has some disadvantages. The blue dye itself can cause adverse effects which include mild skin rash and itching, blue skin discolouration, transient hypotension, bronchospasm and anaphylaxis. The overall risk of developing an adverse reaction to the blue dye is in the region of 0.9% with patent blue and 1.6% with isosulfan blue (Montgomery et al. 2002, Barthelmes et al. 2010).

With future developments, intraoperative imaging that extends beyond the naked eye and allows an accurate assessment of the extent of metastatic disease may become available. This would avoid the need for blue dye administration and the subsequent risks associated with it. There is still work to be done to address the issue of atypical draining patterns in certain individuals. Lymphatic mapping assumes orderly progression of tumour spread to the regional node and biopsy of the first nodes in the lymphatic chain; however, lymphatic interconnections are variable between individuals, and therefore, results can be difficult to interpret. The further development of intraoperative imaging could help in these individuals.

A very high sensitivity is already achievable, setting a benchmark for any new/further development in terms of intraoperative imaging.

5.4.4 Neoadjuvant Systemic Therapy

The arrival of neoadjuvant therapies has allowed previously inoperable breast cancers to be treated with surgery. Neoadjuvant therapy refers to a preoperative systemic treatment given before the definitive treatment and, in the case of breast cancer, surgery. It can include chemotherapy, radiotherapy and hormone modulators. The aim in breast cancer is to downstage a previously inoperable cancer so that it is amenable to surgery and potentially curable treatment. In some cases, it also allows breast conservation surgery as opposed to a mastectomy.

Neoadjuvant treatment has been shown to have an impact on the effectiveness of SLNB. The Sentinel Neoadjuvant (SENTINA) trial was a multicentre study that investigated SLNB before and after neoadjuvant chemotherapy (Kuehn et al. 2013). It discovered that SLNB has a higher detection rate and lower false-negative rate before neoadjuvant chemotherapy. It found that in cases of SLNB performed after neoadjuvant treatments, the detection rate decreased to 80.1% (from 99.1%) and the false-negative rate increased to 14.2%. This has implications for patients undergoing neoadjuvant treatments and highlights the need for improved technical developments in SLNB in this group.

5.5 SUMMARY OF KEY POINTS

1. SLNB started to replace ALND as the first-line procedure for axillary staging during the 1990s.

2. SLNB has lower complications than ALND but with comparable diagnostic results.

3. A dual-technique SLNB, using radioisotope and blue dye, increases specificity and sensitivity.

4. Future developments should focus on improved intraoperative imaging and eliminating the need for blue dye whilst retaining high specificity.

5.6 CONCLUSION

The importance of improving the detection of positive sentinel nodes in breast cancer is obvious. The use of gamma camera technology could vastly improve the accuracy of detection of axillary disease whilst limiting the effects of dual technique SLNB and has the potential to allow intraoperative axillary staging. It also has the potential to be applied to other clinical settings.

REFERENCES

Barthelmes, L., A. Goyal, R.G. Newcombe, F. McNeill and R.E. Mansel. 2010. Adverse reactions to patent blue V dye – The NEW START and ALMANAC experience. *European Journal of Surgical Oncology (EJSO),* 36(4):399–403.

Cabanas, R.M. 1977. An approach for the treatment of penile carcinoma. *Cancer,* 39(2):456–466.

Chetty, U., W. Jack, R.J. Prescott, C. Tyler and A. Rodger. 2000. Management of the axilla in operable breast cancer treated by breast conservation: A randomized clinical trial. *British Journal of Surgery,* 87(2):163–169.

Chiappa, S., C. Uslenghi, G. Bonadonna, P. Marano and G. Ravasi. 1966. Combined testicular and foot lymphangiography in testicular carcinomas. *Surgery, Gynecology & Obstetrics,* 123(1):10–14.

Christensen, B., M. Blichert-Toft, O.J. Siemssen and S.L. Nielsen. 1980. Reliability of axillary lymph node scintiphotography in suspected carcinoma of the breast. *British Journal of Surgery,* 67(9):667–668.

D'Angelo-Donovan, D.D., D. Dickson-Witmer and N.J. Petrelli. 2012. Sentinel lymph node biopsy in breast cancer: A history and current clinical recommendations. *Surgical Oncology,* 21(3):196–200.

Donker, M., G. van Tienhoven, M.E. Straver et al. 2014. Radiotherapy or surgery of the axilla after a positive sentinel node in breast cancer (EORTC 10981–22023 AMAROS): A randomised, multicentre, open-label, phase 3 non-inferiority trial. *The Lancet Oncology*, 15(12):1303–1310.

Giammarile, F., N. Alazraki, J.N. Aarsvold et al. 2013. The EANM and SNMMI practice guideline for lymphoscintigraphy and sentinel node localization in breast cancer. *European Journal of Nuclear Medicine and Molecular Imaging*, 40(12):1932–1947.

Giuliano, A.E., K.K. Hunt, K.V. Ballman et al. 2011. Axillary dissection vs no axillary dissection in women with invasive breast cancer and sentinel node metastasis: A randomized clinical trial. *Journal of the American Medical Association*, 305(6):569–575.

Giuliano, A.E., D.M. Kirgan, J.M. Guenther and D.L. Morton. 1994. Lymphatic mapping and sentinel lymphadenectomy for breast cancer. *Annals of Surgery*, 220(3):391.

Glechner, A., A. Wöckel, G. Gartlehner et al. 2013. Sentinel lymph node dissection only versus complete axillary lymph node dissection in early invasive breast cancer: A systematic review and meta-analysis. *European Journal of Cancer*, 49(4):812–825.

Gould, E.A., T. Winship, P.H. Philbin and H.H. Kerr. 1960. Observations on a "sentinel node" in cancer of the parotid. *Cancer*, 13(1):77–78.

Goyal, A., R.G. Newcombe, A. Chhabra and R.E. Mansel. 2006. Factors affecting failed localisation and false-negative rates of sentinel node biopsy in breast cancer – Results of the ALMANAC validation phase. *Breast Cancer Research and Treatment*, 99(2):203–208.

Haagensen, C.D. 1972. *The Lymphatics in Cancer*. WB Saunders Company, Philadelphia, PA, pp. 300–387.

Huxley, N., T. Jones-Hughes, H. Coelho et al. 2015. A systematic review and economic evaluation of intraoperative tests [RD-100i one-step nucleic acid amplification (OSNA) system and Metasin test] for detecting sentinel lymph node metastases in breast cancer. *Health Technology Assessment*, 19(2):1–215.

Krag, D.N., S.J. Anderson, T.B. Julian et al. 2007. Technical outcomes of sentinel-lymph-node resection and conventional axillary-lymph-node dissection in patients with clinically node-negative breast cancer: Results from the NSABP B-32 randomised phase III trial. *The Lancet Oncology*, 8(10):881–888.

Kuehn, T., I. Bauerfeind, T. Fehm et al. 2013. Sentinel-lymph-node biopsy in patients with breast cancer before and after neoadjuvant chemotherapy (SENTINA): A prospective, multicentre cohort study. *The Lancet Oncology*, 14(7):609–618.

Lyman, G.H., S. Temin, S.B. Edge et al. 2014. Sentinel lymph node biopsy for patients with early-stage breast cancer: American Society of Clinical Oncology clinical practice guideline update. *Journal of Clinical Oncology*, 32(13):1365–1383.

Macaskill, E.J., S. Dewar, C.A. Purdie, K. Brauer, L. Baker and D.C. Brown. 2012. Sentinel node biopsy in breast cancer has a greater node positivity rate than axillary node sample: Results from a retrospective analysis. *European Journal of Surgical Oncology*, 38(8):662–669.

Mansel, R.E., L. Fallowfield, M. Kissin et al. 2006. Randomized multicenter trial of sentinel node biopsy versus standard axillary treatment in operable breast cancer: The ALMANAC Trial. *Journal of the National Cancer Institute*, 98(9):599–609.

Mansel, R.E., F. MacNeill, K. Horgan et al. 2013. Results of a national training programme in sentinel lymph node biopsy for breast cancer. *British Journal of Surgery*, 100(5):654–661.

McMasters, K.M., T.M. Tuttle, D.J. Carlson et al. 2000. Sentinel lymph node biopsy for breast cancer: A suitable alternative to routine axillary dissection in multi-institutional practice when optimal technique is used. *Journal of Clinical Oncology*, 18(13):2560–2566.

Montgomery, L.L., A.C. Thorne, K.J. Van Zee et al. 2002. Isosulfan blue dye reactions during sentinel lymph node mapping for breast cancer. *Anesthesia & Analgesia*, 95(2):385–388.

Morton, D.L., D.R. Wen and A.J. Cochran. October 1989. Pathophysiology of regional lymph node metastases in early melanoma studied by intraoperative mapping of the cutaneous lymphatics. In *Second International Conference on Melanoma*, Venice, Italy, Vol. 131.

Morton, D.L., D.R. Wen, J.H. Wong et al. 1992. Technical details of intraoperative lymphatic mapping for early stage melanoma. *Archives of Surgery*, 127(4):392–399.

Tanis, P.J., O.E. Nieweg, R.A. Valdés Olmos, J. Emiel and B.B. Kroon. 2001. History of sentinel node and validation of the technique. *Breast Cancer Research*, 3(2):109.

Veronesi, U., G. Paganelli, G. Viale et al. 2003. A randomized comparison of sentinel-node biopsy with routine axillary dissection in breast cancer. *New England Journal of Medicine*, 349(6):546–553.

Yarnold, J. 2009. Early and locally advanced breast cancer: Diagnosis and treatment National Institute for Health and Clinical Excellence guideline 2009. *Clinical Oncology*, 21(3):159–160.

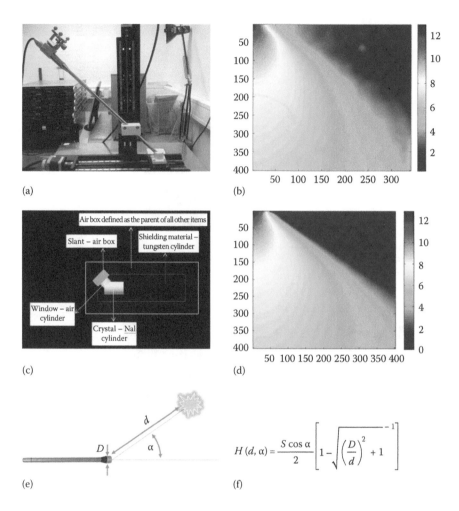

(a)

(b)

(c)

(d)

(e)

(f)

$$H(d, \alpha) = \frac{S \cos \alpha}{2} \left[1 - \left[\sqrt{\left(\frac{D}{d} \right)^2 + 1} \right]^{-1} \right]$$

FIGURE 4.4 Different options for determining the response function of a nuclear detector (here a single-pixel detector). (a) Gamma probe fixed on a positioning table in order to move it relative to a point source and acquire a position by position response. (b) Response function acquired, for example, in (a), as a plane across the detector axis. (c) Simulation model for a gamma probe. (d) Response function simulated, for example, in (c), in a plane across detector. Here the point source was virtually moved in front of the detector. (e) Schematic for simple analytical model for response function. (f) Exemplary simple model based on (e) for response function of a single pixel cylindrical with diameter D and normalised sensitivity S.

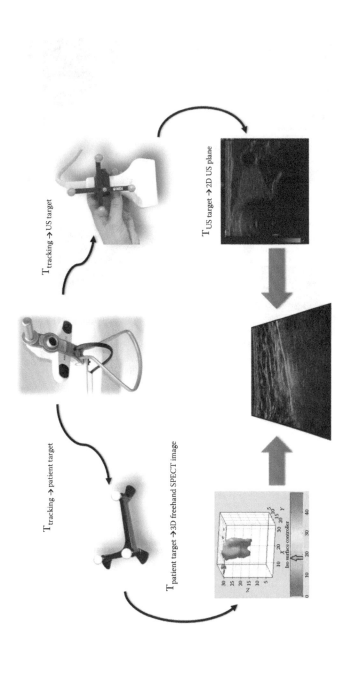

FIGURE 4.7   Image fusion of freehand single-photon emission computed tomography (SPECT) with ultrasound (US). An optical tracking system is used to track both the patient and the US probe using tracking targets. A previously acquired freehand SPECT image is already in the patient target coordinates. By translating and rotating it following the transformation chain shown here, the freehand SPECT image can be put in the coordinate system of the tracking target of the US. Then using the projection matrix that is obtained during the calibration, the freehand SPECT image can be overlaid on the US.

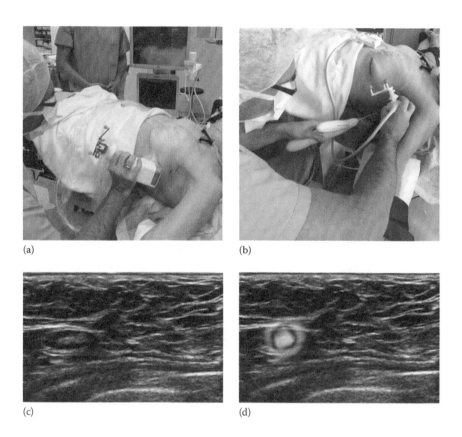

(a) (b)

(c) (d)

FIGURE 4.9   Freehand single-photon emission computed tomography (SPECT)/
ultrasound (US) in action during sentinel lymph node (SLN) aspiration biopsy.
(a) Freehand SPECT acquisition using a handheld gamma camera as nuclear
detector, here in a breast cancer patient. (b) Placement of the needle for aspiration
biopsy based on freehand SPECT/US images of (d). (c) B-mode image of axilla of
patient showing at least one lymph node. (d) Freehand SPECT/US combination
highlights the radioactive SLN by making it more prominent.

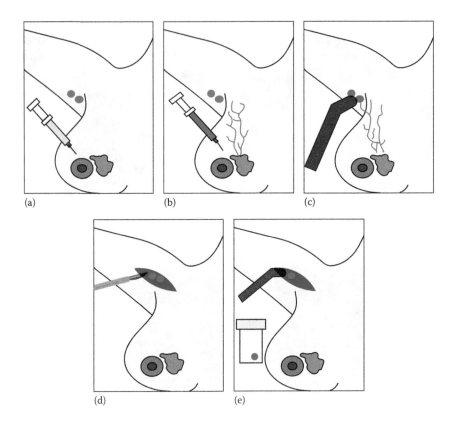

FIGURE 5.2 The procedure of sentinel lymph node biopsy. (a) Radioisotope is injected into the periareolar region before the operation. (b) Blue dye is injected into the periareolar region on the day of SLNB and spreads via lymphatics to the axilla. (c) Gamma probe is used to detect radioactivity uptake in the axilla. (d) Using the gamma probe as a guide, an incison is made. Blue nodes can be seen. (e) The gamma probe helps to identify deeper sentinel nodes and these are removed for analysis.

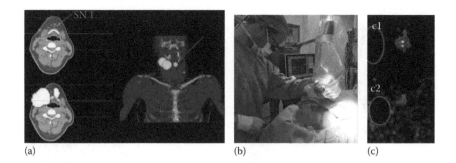

(a)   (b)   (c)

FIGURE 7.4   Head and neck sentinel node (SN) procedure. A 54-year-old patient with a melanoma in the right side of the neck. (a) 3D volume rendering of the SPECT/CT scan. The injection site is clearly visible and an SN at level Ia and a contralateral SN in level IV left. (b) Intraoperative imaging with a PGC, the PGC is aimed at the SN in level Ia. (c) Pre- and post-excision images with the PGC. The blackout zone used in image c1 and c2 is to shield the remaining radiation from the location of the injection site.

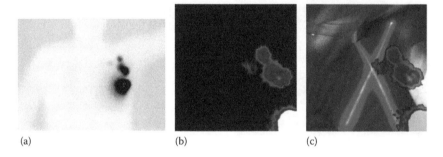

(a)   (b)   (c)

FIGURE 7.6   (a) Conventional lymphoscintigraphy, an injection site with a cluster of sentinel node cranial in this image located in the axilla. (b) Scintigraphic image acquired with a portable gamma camera. (c) Example of a hybrid image. In this image the scintigraphic image and the optical image are displayed in a single view.

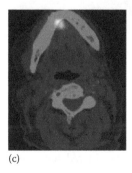

(c)

FIGURE 8.2   Shine-through effect. (c) SPECT/CT showing injection site only, no SN were seen. (Images courtesy of M. McGurk and C. Schilling; Adapted from Clinical and Translational Imaging.)

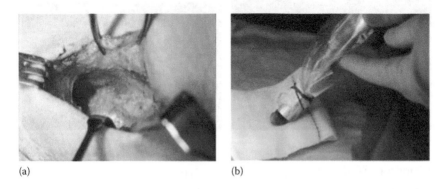

(a)   (b)

FIGURE 8.3   (a) Photograph of a lymph node stained with blue dye as seen during surgery. (b) Ex vivo recording gamma counts using a handheld gamma probe. (Courtesy of M. McGurk.)

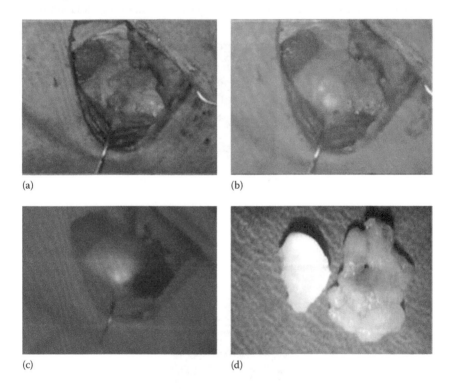

(a)   (b)

(c)   (d)

FIGURE 8.4   Hybrid ICG and Tc^{99m}-nanocolloid tracer for sentinel node biopsy. (a) Access to sentinel node in level II, sternocleidomastoid muscle (SCM) retracted. (b) Surgical field illuminated with near-infrared region (NIR) and white light showing sentinel node (SN) in front of SCM. (c) Surgical field with NIR light only showing fluorescent SN. (d) Fluorescent signal from sentinel node (left) compared to fibroadipose tissue (right) under NIR light. (Courtesy of M. McGurk and C. Schilling.)

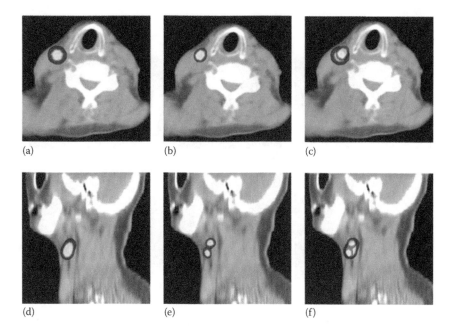

FIGURE 8.6 (a–f) Comparison of SPECT/CT (Tc^{99m}-nanocolloid, red) and PET/CT lymphoscintigraphy (^{89}Zr-nanocolloid, blue). (Adapted from Heuveling, D.A. et al., *J. Nucl. Med.*, 54(4), 585, 2013.)

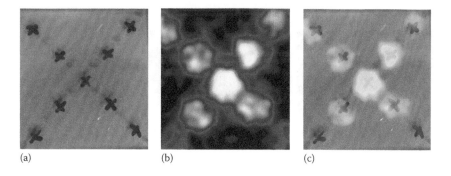

FIGURE 10.3 Illustration of single- and dual-modality hybrid gamma camera images using a bespoke alignment phantom. (a) Optical image, crosses indicate wells containing activity. (b) Gamma image showing the expected nine areas of activity. (c) Hybrid image showing good alignment between the modalities. The phantom contained 13 MBq of ^{99m}Tc and was imaged from a distance of 11 cm. Exposure time was 3.5 min.

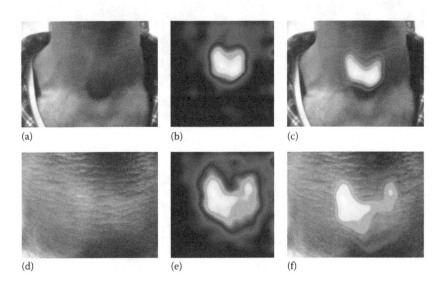

(a)   (b)   (c)

(d)   (e)   (f)

FIGURE 10.8   (a–c) Optical, gamma and combined anterior images of the neck at ~17 cm (300 s image acquisition) starting at 149 min post injection. (d–f) Optical, gamma and combined anterior images of the neck at ~8 cm (300 s image acquisition) starting at 142 min post injection. Gamma images are shown with a 5 pixel width smoothing filter applied.

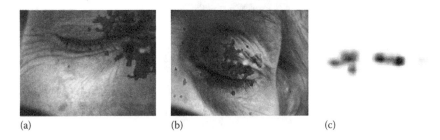

(a)   (b)   (c)

FIGURE 10.9   Hybrid lacrimal drainage scan from the HGC system (a, b) and conventional large field of view (LFOV) lacrimal drainage scan. Hybrid gamma camera (HGC) images taken 1 h 35 min (a) and 1 h 42 min (b) after administration and the conventional LFOV lacrimal drainage scan (c). HGC imaging was at a distance of 7 cm with an exposure time of 5 min.

CHAPTER 6

Radiopharmaceuticals for Intraoperative Imaging

J.R. Ballinger

CONTENTS

6.1 INTRODUCTION

The majority of radiopharmaceuticals currently used for intraoperative imaging were not specifically developed for that purpose. As will be shown, the choice of agent has mainly been determined by availability rather than optimised properties. However, the last few years have seen the development of improved agents and expansion into multimodal, hybrid detection.

Most intraoperative imaging methods have involved localisation of lymph nodes (Bricou et al. 2013, Vidal-Sicart et al. 2014). Nuclear imaging of the lymphatic system and lymph nodes has traditionally been performed with radiolabelled particles or relatively large molecules which are administered as an interstitial depot and then migrate via lymphatic channels and are retained to a greater or lesser extent in lymph nodes (Uren et al. 1993). In general, only a small proportion of the radioactivity leaves the injection depot, which can create problems when imaging lymph nodes adjacent to the depot.

The most widely used radionuclide in nuclear medicine is technetium-99m (^{99m}Tc), which has a half-life of 6 h and emits gamma photons with an energy of 140 keV. It is conveniently available from an on-site generator loaded with molybdenum-99 (^{99}Mo), which decays to produce ^{99m}Tc. Other radionuclides which will be discussed include indium-111

TABLE 6.1 ^{99m}Tc-Labelled Radiopharmaceuticals for Intraoperative Imaging

Classification	Subclassification	Radiopharmaceutical	Particle Size Range or Molecular Weight
Particulate agents	Inorganic	Sulphide colloid (SC)	100–1,000 nm
		Filtered sulphide colloid (FSC)	<100 nm
		Rhenium sulphide colloid (ReSC)	50–200 nm
		Antimony sulphide colloid (SbSC)	3–40 nm
		Tin colloid (TC)	30–250 nm
	Albumin based	Nanocolloidal albumin (NC)	7–15 nm
		Macroaggregated albumin (MAA)	10–100 μm
Soluble agents	Non-specific	Human serum albumin (HSA)	67,000 Da
		Immunoglobulin G (IgG)	150,000 Da
	Targeted	Tilmanocept	~20,000 Da
		Monoclonal antibodies (mAb)	150,000 Da
		Sestamibi (MIBI)	778 Da
		^{18}F-fluorodeoxyglucose (FDG)	181 Da
Multimodal agents	Fluorescent radioactive	Indocyanin green-^{99m}Tc-nanocolloid (ICG-NC)	~10 nm
		GE-137	~3,000 Da
	Magnetic radioactive	Superparamagnetic iron oxide (SPIO) particles	60 nm

(^{111}In, t½ 67 h, γ energies 171 and 245 keV), fluorine-18 (^{18}F, t½ 110 min, positron emitter) and gallium-68 (^{68}Ga, t½ 68 min, positron emitter).

The following discussion of the range of radiopharmaceuticals used for intraoperative imaging will be arranged according to the physico-chemical properties of the agents, as summarised in Table 6.1.

6.2 PARTICULATE AGENTS

Most intraoperative imaging methods at present involve the visualisation of a lymph node, in particular the so-called sentinel lymph node (Uren et al. 1993). Adoption of the sentinel node technique has revolutionised surgery in a range of cancers, reducing morbidity without compromising outcomes. The standard sentinel node technique, which involves preoperative imaging and intraoperative localisation with a handheld detector, developed from lymphoscintigraphy. The most widely used agents for lymphoscintigraphy are ^{99m}Tc labelled colloidal particles which are injected interstitially as a depot and which migrate via lymphatic channels and are trapped non-specifically in lymph nodes, primarily due to phagocytosis by monocytes (Mariani et al. 2001).

The optimal size of a colloidal particle for lymphoscintigraphy is a trade-off between competing factors (Mariani et al. 2001). Small particles (<5 nm in diameter) are liable to penetrate capillaries and be carried away in blood, while larger particles (>400 nm) will not be carried by lymph nodes and will remain in the injection depot. Between those extremes, there is not absolute agreement on the optimal particle size. The upper portion of the range (100–400 nm) is suggested by some authors to be optimal because the radiocolloid should be trapped only in first echelon nodes (De Cicco et al. 1998), though this is at the expense of slower and less extensive migration due to size. In contrast, the lower portion of the range (5–100 nm) would allow more rapid delivery but possibly extend to the second echelon nodes and potentially cause overestimation of the number of sentinel nodes.

This is further complicated by the limitations in particle sizing techniques. A variety of particle sizing techniques have been used, including transmission electron microscopy (EM), scanning EM, light microscopy, filtration, centrifugation, Coulter counter, photon correlation spectroscopy and gel chromatography. Many of these measure different parameters and are not directly comparable (Bergqvist et al. 1983, Bergqvist and Strand 1989). Furthermore, there can be a discrepancy between physical measurements of particle size and measurements of the distribution of radioactivity. Larger particles have a greater surface area, and thus, more atoms of ^{99m}Tc can attach to a single particle. Thus, radioactivity measurements using filtration techniques will give an apparently larger mean size, although this may actually be the most useful parameter in practice.

There is also controversy about the role of specific radioactivity (radioactivity per gram or per particle), which determines the number of particles administered. Is it preferable to have the desired quantity of radioactivity delivered in a small number of particles to avoid saturating the potentially limited number of target sites (such as macrophages in the lymph node), or should there be a large number of particles to efficiently deliver the signal to the node? It can be argued that if radiocolloid particles leave the injection depot by a zero-order process (i.e. in a linear manner) then higher average activity per particle should deliver higher nodal counts. Gommans et al. (2001) and Krynyckyi et al. (2002) presented evidence to support this. On the other hand, Valdes Olmos et al. (2001) and Bourgeois (2007) found the opposite effect. Clinical experience suggests that high specific activity may be desirable but this is not a critical parameter (Chakera et al. 2009, Giammarile et al. 2014).

The choice of agent is determined by availability rather than optimal properties and this results in international differences. Thus, the main agents used are ^{99m}Tc-nanocolloid (NC) in Europe, ^{99m}Tc-sulphide colloid (SC) in the United States, ^{99m}Tc-tin colloid in Japan and ^{99m}Tc-antimony SC in Australia (Chakera et al. 2009).

6.2.1 Inorganic Particulate Agents

The first radiopharmaceutical to be widely used for lymphoscintigraphy was ^{99m}Tc-SC. Originally developed to image the reticuloendothelial system (RES) following intravenous administration, ^{99m}Tc-SC is formed by heating a solution of ^{99m}Tc-pertechnetate and sodium thiosulphate at acidic pH in a boiling water bath for 3–5 min, followed by neutralisation by addition of a buffer (Szymendera et al. 1971). In some formulations, perrhenate is added as a carrier. This is one of the few ^{99m}Tc-labelled radiopharmaceuticals in which Tc is present in its original +7 oxidation state and the formula is believed to be Tc_2S_7. A wide range of particles are formed by this method, with diameters of 100–1000 nm (Bergqvist et al. 1983). This range can be reduced by passing the suspension through a 100 or 200 nm membrane filter in order to remove the larger colloidal particles (Hung et al. 1995).

Rhenium sulphide colloid (ReSC, Nanocis®) is prepared in a different manner. Preformed colloidal particles are labelled with ^{99m}Tc by heating with a reducing agent (Bensimhon et al. 2008). Because the colloid is preformed, rather than growing during heating as with SC, the particle range is narrower, 50–200 nm (Bergqvist et al. 1983). This product is mainly used in France.

Smaller colloidal particles with an even narrower range (3–40 nm) can be obtained by substituting antimony trisulphide for sodium thiosulphate, producing ^{99m}Tc-antimony (Sb; SC) (Ege and Warbick 1979, Bergqvist et al. 1983). Although ^{99m}Tc-SbSC has superior properties for lymphoscintigraphy, it is commercially available only in Australia and thus has limited use elsewhere.

Another RES agent occasionally used for lymphoscintigraphy is ^{99m}Tc-tin colloid (TC, Hepatate®). ^{99m}Tc is reduced with stannous fluoride, forming particles with a range of 30–250 nm (Jimenez et al. 2008). A modified formulation is available in Japan where it is widely used for lymphoscintigraphy (Sato et al. 2004).

6.2.2 Albumin-Based Particulate Agents

^{99m}Tc-labelled nanocolloidal human serum albumin (HSA; NC nanocolloid, [NC]) was developed for inflammation imaging but quickly became used for lymphoscintigraphy (Ballinger 2015). HSA is heat denatured in the presence of stannous chloride, then lyophilised. It is stated that >95% of the particles are <80 nm in diameter, though measurements suggest a relatively narrow range of 7–15 nm (Gommans et al. 2001). ^{99m}Tc-NC forms when the lyophilised powder is reconstituted with ^{99m}Tc-pertechnetate solution. In addition to the original product (Nanocoll®, Sorin/GE Healthcare) there are now two generic products licensed in Europe with the same specifications (Nanotop®, Rotop; Nanoscan, Medi-Radiopharma) and another with a larger particle size range of 100–600 nm (Sentiscint®, Medi-Radiopharma). NC is recommended in most of the European procedure guidelines (Alkureishi et al. 2009, Chakera et al. 2009).

An alternative nuclear medicine technique that does not use lymphoscintigraphy is radioguided occult lesion localisation (ROLL), where larger particles of the lung perfusion imaging agent ^{99m}Tc-macroaggregated albumin, 10–100 μm in diameter, are injected under ultrasound guidance into non-palpable tumours. Unlike lymphoscintigraphy, the particles do not migrate but remain at the injection site so that during surgery the surgeon can use a handheld probe or portable gamma camera to locate the tumour. This technique has demonstrated advantages over wire guided localisation (Sajid et al. 2012). More recently, the technique has been rebranded as sentinel node and occult lesion localisation (SNOLL) with ^{99m}Tc-NC as radiotracer (Bricou et al. 2013, Bluemel et al. 2015).

6.3 SOLUBLE AGENTS

Soluble agents that are retained or at least delayed in lymph nodes can be used for lymphoscintigraphy following the interstitial injection. These may be non-specific, with only their large molecular size leading to lymph node visualisation, or they may be targeting features of the lymph node or infiltrating tumour cells.

6.3.1 Non-Specific Soluble Agents

HSA (non-denatured) has been labelled with ^{99m}Tc and used for lymphatic studies (Nathanson et al. 1996, Svensson et al. 1999). An attempt to extend this to labelling with ^{111}In failed due to instability of the label in vivo which was not evident in preclinical testing (Pain et al. 2002). ^{99m}Tc-labelled polyclonal immunoglobulin G has also been used for this purpose (Svensson et al. 1999), though ^{99m}Tc-human immunoglobulin (Technescan HIG, Mallinckrodt) has now been withdrawn from the market for commercial reasons. HIG showed better delineation of lymph nodes than HSA, which was suggested to be due to Fc receptor mediated binding (Svensson et al. 1999). HIG has also been labelled with ^{111}In via a modified DPTA chelator for lymphatic studies (Pain et al. 2002, Peters et al. 2009). The plasma volume expanders dextran and hydroxyethyl starch have been complexed with ^{99m}Tc for lymphoscintigraphy and widely used in countries such as Brazil (Masiero et al. 2005), but there have been no commercially available products internationally.

6.3.2 Targeted Soluble Agents

Recently, tilmanocept (LymphoSeek®, Navidea) has been licensed for clinical use in the United States and Europe. It is the first agent specifically designed for lymph node detection. It consists of a dextran backbone that is derivatised with mannose side chains, designed to interact with the CD206 mannose receptor on macrophages in lymph nodes, and DTPA chelation sites for stable binding to ^{99m}Tc (Ellner et al. 2003, Sondak et al. 2013). As a soluble molecule of moderate size, it can leave the depot efficiently and migrate along lymphatic channels to lymph nodes. Although the fraction of the injected activity that reaches nodes is not significantly higher than that observed with colloids, image quality is superior because of greater migration from the injection depot. However, as a new product with substantial development costs to be recouped, it is considerably more expensive than existing agents and this may limit its market penetration unless significant clinical benefits can be demonstrated (Baker et al. 2015). The same targeting approach has been applied to mannosylated-HSA labelled with ^{68}Ga (Eo et al. 2015).

More than 30 years ago, it was proposed to use interstitial injection of labelled monoclonal antibodies to detect tumour cells in lymph nodes (Weinstein et al. 1983). In theory, this should give much greater specificity for detection of lymph node metastases without the background activity due to slow clearance of antibodies from the circulation following intravenous injection. However, any lymph node localisation that has been observed seems to be due to non-specific retention, and the relatively small number of tumour cells in lymph nodes may not allow adequate sensitivity (Bartolomei et al. 1998).

Finally, a number of tumour seeking radiopharmaceuticals show a sufficient retention in primary tumours and lymph nodes to be useful in radioguided surgery. One such radiopharmaceutical is ^{99m}Tc sestamibi, which was originally developed for myocardial perfusion imaging but that is taken up in a range of tumours because of their hyperpolarised mitochondria (Fukumoto 2004). Sestamibi is useful for guiding surgery in parathyroid adenomas and primary hyperparathyroidism (Denmeade et al. 2013, Grassetto and Rubello 2013). Sestamibi is also used in primary breast cancer and its locoregional metastases in a procedure called scintimammography, which has also been incorporated into intraoperative imaging (Evangelista et al. 2014).

Finally in this category, ^{18}F fluorodeoxyglucose can be used for intraoperative imaging because of its intense uptake in primary and metastatic tumours with relatively low levels in normal tissues (Gulec et al. 2007, Piert et al. 2008). More recently, Cerenkov imaging has been used to detect the luminescence generated in tissues by positron emitting radionuclides (Das et al. 2014). High image resolution can be achieved, but only superficial tumours can be detected because of the limited penetration of the photons. The technique has also been applied ex vivo to study tumour margins of excised tissues (Grootendorst and Purushotham 2015).

6.4 MULTIMODAL AGENTS

In the last 10–15 years, we have seen how the addition of computed tomography to single-photon emission computed tomography and positron emission tomography has greatly enhanced the diagnostic utility of the techniques. Similarly, a combination of optical (visual or fluorescent) or magnetic techniques with conventional radiotracer technology is poised to revolutionise minimally invasive surgery (Vidal-Sicart et al. 2015).

6.4.1 Fluorescent-Radioactive Multimodal Agents

Sentinel node localisation has generally involved separate injections of a radiotracer and blue dye. The two cannot be combined because of their differing kinetics. Moreover, the presence of blood in the surgical field makes visualisation of the blue dye difficult. Higher sensitivity and contrast could be obtained by the use of a fluorescent tracer. To this end, van Leeuwen et al. designed indocyanin green-^{99m}Tc-nanocolloid (ICG-NC) as a hybrid tracer that can be conveniently prepared from licensed components (Buckle et al. 2010). ^{99m}Tc-NC is prepared in the standard manner, then a small amount of ICG (a fraction of the amount approved for use as a single agent) is added to the vial and the complex self-assembles virtually instantly. The utility of this hybrid tracer has been demonstrated in a range of cancers (Vidal-Sicart et al. 2015). Because the fluorescence properties of ICG are not ideal, other near infrared dyes are being explored, including a fluorescent analogue of tilmanocept (Hosseini et al. 2014). This topic will be discussed in much greater detail in another chapter in this book.

In parallel with what was mentioned earlier, there has been progress in the development of fluorescent agents for delineation of tumour margins. One such agent is GE-137, a fluorescence-tagged 26 amino acid cyclic peptide which binds the human tyrosine kinase c-Met (Burggraaf et al. 2015).

6.4.2 Magnetic-Radioactive Multimodal Agents

The use of radiolabelled probes is associated with a variety of challenges: logistics of supply due to radioactive half-life; radiation protection concerns before, during and after surgery; and regulatory issues. Non-radioactive alternatives are under investigation. One approach which shows some promise is the use of superparamagnetic iron oxide (SPIO) particles detected by a handheld magnetometer. This has been shown to produce results equivalent to standard techniques in breast cancer (Douek et al. 2014). However, validation of the technique will require magnetic-radioactive multimodal agents which are still under development. The SPIO approach has also been evaluated with the ROLL technique described earlier (Ahmed et al. 2015).

6.5 SUMMARY

A variety of radiopharmaceuticals have proven useful for intraoperative imaging, primarily for the localisation of lymph nodes. Most of the radiotracers were developed and licensed for other clinical indications, but their commercial availability allows them to be used for intraoperative imaging. The range of radiotracers varies between countries for commercial reasons. Recently, we have seen the development of agents specifically for intraoperative imaging, and this trend will likely continue with multimodal agents which can make full use of the complementary information from optical and radiotracers techniques.

REFERENCES

Ahmed, M., B. Anninga, S. Goyal, P. Young, Q.A. Pankhurst and M. Douek. 2015. Magnetic sentinel node and occult lesion localization in breast cancer (MagSNOLL Trial). *British Journal of Surgery*, 102(6):646–652.

Alkureishi, L.W., Z. Burak, J.A. Alvarez et al. 2009. Joint practice guidelines for radionuclide lymphoscintigraphy for sentinel node localization in oral/oropharyngeal squamous cell carcinoma. *Annals of Surgical Oncology*, 16(11):3190–3210.

Baker, J.L., M. Pu, C.A. Tokin et al. 2015. Comparison of (^{99m}Tc) tilmanocept and filtered (^{99m}Tc) sulfur colloid for identification of SLNs in breast cancer patients. *Annals of Surgical Oncology*, 22(1):40–45.

Ballinger, J.R. 2015. The use of protein-based radiocolloids in sentinel node localisation. *Clinical and Translational Imaging*, 3(3):179–186.

Bartolomei, M., A. Testori, M. Chinol et al. 1998. Sentinel node localization in cutaneous melanoma: Lymphoscintigraphy with colloids and antibody fragments versus blue dye mapping. *European Journal of Nuclear Medicine and Molecular Imaging*, 25(11):1489–1494.

Bensimhon, L., T. Metaye, J. Guilhot and R. Perdrisot. 2008. Influence of temperature on the radiochemical purity of ^{99m}Tc-colloidal rhenium sulfide for use in sentinel node localization. *Nuclear Medicine Communications*, 29(11):1015–1020.

Bergqvist, L. and S.E. Strand. 1989. Autocorrelation spectroscopy for particle sizing and stability tests of radiolabelled colloids. *European Journal of Nuclear Medicine and Molecular Imaging*, 15(10):641–645.

Bergqvist, L., S.E. Strand and B.R. Persson, 1983, January. Particle sizing and biokinetics of interstitial lymphoscintigraphic agents. *Seminars in Nuclear Medicine*, 13(1):9–19.

Bluemel, C., A. Cramer, C. Grossmann et al. 2015. iROLL: Does 3-D radioguided occult lesion localization improve surgical management in early-
stage breast cancer? *European Journal of Nuclear Medicine and Molecular
Imaging, 42*(11):1692–1699.

Bourgeois, P. 2007. Scintigraphic investigations of the lymphatic system: The
influence of injected volume and quantity of labeled colloidal tracer. *Journal
of Nuclear Medicine, 48*(5):693–695.

Bricou, A., M.A. Duval, Y. Charon and E. Barranger. 2013. Mobile gamma cameras in breast cancer care – A review. *European Journal of Surgical Oncology,
39*(5):409–416.

Buckle, T., A.C. Van Leeuwen, P.T. Chin et al. 2010. A self-assembled multimodal
complex for combined pre-and intraoperative imaging of the sentinel
lymph node. *Nanotechnology, 21*(35):355101.

Burggraaf, J., I.M. Kamerling, P.B. Gordon et al. 2015. Detection of colorectal
polyps in humans using an intravenously administered fluorescent peptide
targeted against c-Met. *Nature Medicine, 21*(8):955–961.

Chakera, A.H., B. Hesse, Z. Burak et al. 2009. EANM-EORTC general recommendations for sentinel node diagnostics in melanoma. *European Journal
of Nuclear Medicine and Molecular Imaging, 36*(10):1713–1742.

Das, S., D.L. Thorek, and J. Grimm. 2014. Cerenkov imaging. *Advances in Cancer
Research* 124:213–234.

De Cicco, C., M. Cremonesi, A. Luini et al. 1998. Lymphoscintigraphy and
radioguided biopsy of the sentinel axillary node in breast cancer. *Journal of
Nuclear Medicine, 39*:2080–2083.

Denmeade, K.A., C. Constable and W.M. Reed. 2013. Use of ^{99m}Tc 2-methoxyiso-
butyl isonitrile in minimally invasive radioguided surgery in patients with
primary hyperparathyroidism: A narrative review of the current literature.
Journal of Medical Radiation Sciences, 60(2):58–66.

Douek, M., J. Klaase, I. Monypenny et al. 2014. Sentinel node biopsy using a magnetic tracer versus standard technique: The SentiMAG multicentre trial.
Annals of Surgical Oncology, 21(4):1237–1245.

Ege, G.N. and A. Warbick. 1979. Lymphoscintigraphy: A comparison of ^{99}Tcm
antimony sulphide colloid and ^{99}Tcm stannous phytate. *British Journal of
Radiology, 52*(614):124–129.

Ellner, S.J., C.K. Hoh, D.R. Vera, D.D. Darrah, G. Schulteis and A.M. Wallace.
2003. Dose-dependent biodistribution of (^{99m}Tc) DTPA-mannosyl-dextran
for breast cancer sentinel lymph node mapping. *Nuclear Medicine and
Biology, 30*(8):805–810.

Eo, J.S., H.K. Kim, S. Kim, Y.S. Lee, J.M. Jeong and Y.H. Choi. 2015. Gallium-68
neomannosylated human serum albumin-based PET/CT lymphoscintigraphy for sentinel lymph node mapping in non-small cell lung cancer. *Annals
of Surgical Oncology, 22*(2):636–641.

Evangelista, L., A.R. Cervino, R. Sanco et al. 2014. Use of a portable gamma camera for guiding surgical treatment in locally advanced breast cancer in a
post-neoadjuvant therapy setting. *Breast Cancer Research and Treatment,
146*(2):331–340.

Fukumoto, M. 2004. Single-photon agents for tumor imaging: 201TI, ^{99m}Tc-MIBI, and ^{99m}Tc-tetrofosmin. *Annals of Nuclear Medicine, 18*(2):79–95.

Giammarile, F., M.F. Bozkurt, D. Cibula et al. 2014. The EANM clinical and technical guidelines for lymphoscintigraphy and sentinel node localization in gynaecological cancers. *European Journal of Nuclear Medicine and Molecular Imaging, 41*(7):1463–1477.

Gommans, G.M., A. van Dongen, T.G. van der Schors et al. 2001. Further optimisation of ^{99m}Tc-Nanocoll sentinel node localisation in carcinoma of the breast by improved labelling. *European Journal of Nuclear Medicine and Molecular Imaging, 28*:1450–1455.

Grassetto, G. and D. Rubello. 2013. The increasing role of minimal invasive radioguided parathyroidectomy for treating single parathyroid adenoma. *Journal of Postgraduate Medicine, 59*(1):1.

Grootendorst, M.R. and A. Purushotham. 2015. Clinical feasibility of intraoperative ^{18}F-FDG Cerenkov Luminescence Imaging in breast cancer surgery. *Journal of Nuclear Medicine, 56*(3):13.

Gulec, S.A., E. Hoenie, R. Hostetter and D. Schwartzentruber. 2007. PET probe-guided surgery: Applications and clinical protocol. *World Journal of Surgical Oncology, 5*(1):65.

Hosseini, A., J.L. Baker, C.A. Tokin et al. 2014. Fluorescent-tilmanocept for tumor margin analysis in the mouse model. *Journal of Surgical Research, 190*(2):528–534.

Hung, J.C., G.A. Wiseman, H.W. Wahner, B.P. Mullan, T.R. Taggart and W.L. Dunn. 1995. Filtered technetium-99m-sulfur colloid evaluated for lymphoscintigraphy. *Journal of Nuclear Medicine, 36*(10):1895–1901.

Jimenez, I.R., M. Roca, E. Vega et al. 2008. Particle sizes of colloids to be used in sentinel lymph node radiolocalization. *Nuclear Medicine Communications, 29*(2):166–172.

Krynyckyi, B.R., Z.Y. Zhang, C.K. Kim, H. Lipszyc, K. Mosci and J. Machac. 2002. Effect of high specific-activity sulfur colloid preparations on sentinel node count rates. *Clinical Nuclear Medicine, 27*(2):92–95.

Mariani, G., L. Moresco, G. Viale et al. 2001. Radioguided sentinel lymph node biopsy in breast cancer surgery. *Journal of Nuclear Medicine, 42*(8):1198–1215.

Masiero, P.R., N.L. Xavier, B.L. Spiro, M.F. Detanico, M.D.C. Xavier and A.L. Pinto. 2005. Scintigraphic sentinel node detection in breast cancer patients: Paired and blinded comparison of ^{99m}Tc dextran 500 and ^{99m}Tc phytate. *Nuclear Medicine Communications, 26*(12):1087–1091.

Nathanson, S.D., L. Nelson and K.C. Karvelis. 1996. Rates of flow of technetium 99m-labeled human serum albumin from peripheral injection sites to sentinel lymph nodes. *Annals of Surgical Oncology, 3*(4):329–335.

Pain, S.J., R.S. Nicholas, R.W. Barber et al. 2002. Quantification of lymphatic function for investigation of lymphedema: Depot clearance and rate of appearance of soluble macromolecules in blood. *Journal of Nuclear Medicine, 43*(3):318–324.

Peters, A.M., J.C. Fowler, T.B. Britton et al. 2009. Functional variation in lymph node arrangements within the axilla. *Lymphatic Research and Biology, 7*(3):139–144.

Piert, M., J. Carey and N. Clinthorne. 2008. Probe-guided localization of cancer deposits using [¹⁸F] fluorodeoxyglucose. *Quarterly Journal of Nuclear Medicine and Molecular Imaging*, 52(1):37.

Sajid, M.S., U. Parampalli, Z. Haider and R. Bonomi. 2012. Comparison of radioguided occult lesion localization (ROLL) and wire localization for non-palpable breast cancers: A meta-analysis. *Journal of Surgical Oncology*, 105(8):852–858.

Sato, K., D. Krag, K. Tamaki et al. 2004. Optimal particle size of radiocolloid for sentinel node identification in breast cancer–Electron microscopic study and clinical comparison. *Breast Cancer*, 11(3):256–263.

Sondak, V.K., D.W. King, J.S. Zager et al. 2013. Combined analysis of phase III trials evaluating [99mTc] tilmanocept and vital blue dye for identification of sentinel lymph nodes in clinically node-negative cutaneous melanoma. *Annals of Surgical Oncology*, 20(2):680–688.

Svensson, W., D.M. Glass, D. Bradley and A.M. Peters. 1999. Measurement of lymphatic function with technetium-99m-labelled polyclonal immunoglobulin. *European Journal of Nuclear Medicine and Molecular Imaging*, 26(5):504–510.

Szymendera, J., T. Zoltowski, M. Radwan and J. Kaminska. 1971. Chemical and electron microscope observations of a safe PVP-stabilized colloid for liver and spleen scanning. *Journal of Nuclear Medicine*, 12(5):212–215.

Uren, R.F., R.B. Howman-Giles, H.M. Shaw, J.F. Thompson and W.H. McCarthy. 1993. Lymphoscintigraphy in high-risk melanoma of the trunk: Predicting draining node groups, defining lymphatic channels and locating the sentinel node. *Journal of Nuclear Medicine*, 34:1435–40.

Valdés Olmos, R.A.V., P.J. Tanis, C.A. Hoefnagel et al. 2001. Improved sentinel node visualization in breast cancer by optimizing the colloid particle concentration and tracer dosage. *Nuclear Medicine Communications*, 22(5):579–586.

Vidal-Sicart, S., M.E. Rioja, P. Paredes, M.R. Keshtgar and R.A. Valdés Olmos. 2014. Contribution of perioperative imaging to radioguided surgery. *Quarterly Journal of Nuclear Medicine and Molecular Imaging*, 58:140–160.

Vidal-Sicart, S., F.W. van Leeuwen, N.S. van den Berg and R.A. Valdés Olmos. 2015. Fluorescent radiocolloids: Are hybrid tracers the future for lymphatic mapping? *European Journal of Nuclear Medicine and Molecular Imaging*, 11(42):1627–1630.

Weinstein, J.N., M.A. Steller, A.M. Keenan et al. 1983. Monoclonal antibodies in the lymphatics: Selective delivery to lymph node metastases of a solid tumor. *Science*, 222(4622):423–426.

CHAPTER 7

Surgical Experiences with Intraoperative Gamma Cameras

B. Pouw, L.J. de Wit-van der Veen
and M.P.M. Stokkel

CONTENTS

7.1 CLINICAL VALUE AND RELEVANCE IN CLINICAL ROUTINE

7.1.1 General Background

In current nuclear medicine practice, imaging is routinely performed either with conventional gamma cameras, nowadays commonly equipped with CT scanners, or with PET/CT scanners. In addition to these large devices, several intraoperative small field of view (SFOV) portable gamma cameras (PGCs) have become available over the past years for intraoperative imaging. With these devices, high-resolution images of small surface areas can be obtained and local radioactivity distribution patterns can be assessed with a relative short image acquisition time. A PGC aids the surgeon to localise radioactive targets during surgery, and it can also be used to guide certain interventions, such as sentinel node (SN) mapping and biopsy.

Radioguided surgery is traditionally performed with a gamma probe to assist the surgeon to easily identify a specific target. These commercial systems are compact and easy to use and have a high sensitivity. Nonetheless, probes are collimated to a narrow FOV and only give acoustic feedback on the count rate without offering spatial information. Indeed, PGCs are known for their better spatial resolution (i.e. resolving power or ability to distinguish two hotspots) compared to a gamma probe and provide a real-time overview of the radioactivity distribution in the imaged area. An additional benefit of surgery with a PGC is the ability to record all the consecutive steps in a procedure (Figure 7.1).

7.1.2 Technical Features and Practical Considerations

Soluri and Pani patented the first PGC in 1997, called an imaging probe (Soluri and Pani 2001). This camera had a FOV of 2.54 cm^2. Today, there are several different PGC systems commercially available, such as Sentinella 102 (Oncovision, Valencia, Spain), the CrystalCam (Crystal Photonics GbmH, Berlin, Germany), and soon the *NebulEYE*, Gamma Technologies Ltd in Leicester, United Kingdom (Sánchez et al. 2006, Bugby et al. 2014, Knoll et al. 2014). In addition, many other PGC systems have been developed in the past, which were mainly used in clinical research settings (Abe et al. 2003, Barranger et al. 2008, Bricou et al. 2015).

Based on clinical experiences with these systems, there are a number of practical considerations that have to be taken into account when developing a PGC system:

1. It should have a sensitivity high enough to detect faint accumulation in lymph nodes.

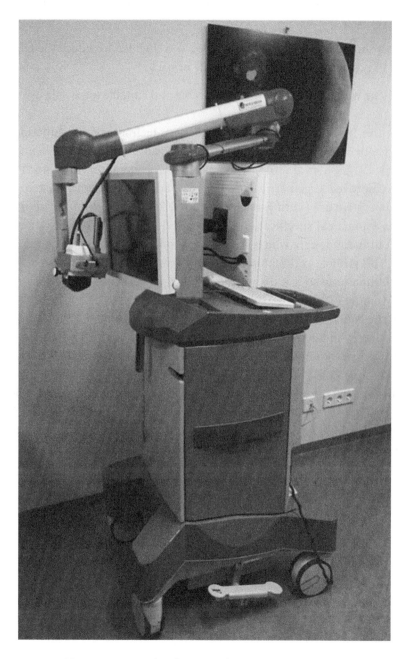

FIGURE 7.1   This is an example of a portable gamma camera (PGC); this PGC has a support system for the camera head for stable image acquisitions.

2. It should have the spatial resolution to distinguish between two targets.

3. It should be adequately collimated and shielded to reduce the effects of radiation coming from the side or back.

4. The whole system should be movable and suitable for sterile use.

Since the 1997 variations in camera design, technology and functionality have been proposed. Initially, handheld gamma cameras of 1 or 2 kg emerged (Russo et al. 2011), followed by the generation of lighter PGC with improved ergometrical details and adequate support system for intraoperative use (Vermeeren et al. 2009a, Knoll et al. 2014). In addition to these practical specifications, the main difference between systems is the collimator design, which is either a pinhole or parallel hole configuration. The pinhole collimator enables a variable FOV, which depends on the collimator to source distance, while the parallel collimator has a fixed FOV (Figure 7.2).

In the Netherlands Cancer Institute, we started using PGC for radioguided surgery in 2002 with a subsequent increased use for various indications (Vermeeren et al. 2009b, Brouwer et al. 2011). The first PCGs were an eZ-Scope (2002, Yokohama City University, Kanagawa, Japan) and a Minicam (2003, Eurorad SA, Eckbolsheim, France) (Otake et al. 2002, Valdés Olmos et al. 2014a). These two models were heavy handheld PCG devices. In 2005, a PCG, Sentinella 102, with improved ergonomic design and adequate support system for intraoperative use was introduced.

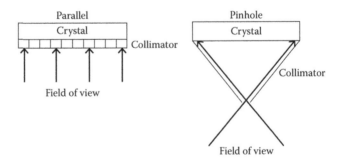

FIGURE 7.2 This illustration shows a schematic overview of a parallel and a pinhole collimator. The parallel collimator on the left side has a fixed field of view (FOV), and the pinhole collimator field of view (FOV) increases with a greater distance because of the diverging of FOV.

This PGC was based on a cesium iodide (sodium; CsI(Na)) continuous scintillating crystal equipped with a 4 mm pinhole collimator (a 2 mm pinhole collimator is also available). The FOV of the pinhole camera is 4 cm^2 when placed at 3 cm from the imaging plane and increases to 20 cm^2 when placed at a distance of 15 cm. The intrinsic spatial resolution is 1.8 mm, while the extrinsic spatial resolution is 7 and 21 mm at distances of 3 and 15 cm, respectively. Detection sensitivity for the 4 mm pinhole collimator depends on the distance to the imaging plane, being 319 cps/MBq and 18 cps/MBq for distances of 3 and 15 cm, respectively. These and other technical details of this PGC are described in more detail by Sánchez et al. (Sánchez et al. 2006, Vermeeren et al. 2009b; Figure 7.3).

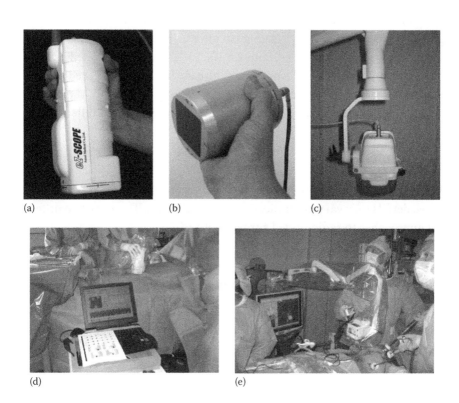

(a)   (b)   (c)

(d)   (e)

FIGURE 7.3 Historical evolution of the use of portable gamma cameras at the Netherlands Cancer Institute: (a) Handheld eZ-Scope camera with a weight of approximately 1 kg (2002); (b) Minicam with a weight of about 2 kg (2003); (c) Sentinella, the first PCG with supportive system (2005); (d) eZ-Scope used for a sentinel node (SN) procedure of the groin (2003); (e) Sentinella camera used for a laparoscopic SN procedure (2006). (Courtesy of Dr. Renato A. Valdés Olmos.)

Most PGCs are suitable for detecting radioisotopes with energies ranging from approximately 30 to 250 keV. Their main application is the detection of ^{99m}Tc-labelled tracers, such as ^{99m}Tc-nanocolloid, with peak radiation energy of 140.5 keV. Although other tracers like ^{111}I, ^{57}Co, ^{153}Gd, ^{123}I and ^{125}I in the form of small implantable seeds can be easily detected, their application is less common.

7.1.3 Clinical Indications

The most frequent application of PGCs in our institute is intraoperative imaging to aid SN mapping and the SN biopsy in head and neck cancer, melanoma, prostate cancer, penile cancer, vulvar cancer, kidney cancer and testicular cancer. In addition, it is also used in our hospital to monitor systemic toxicity during isolated limb chemoperfusion. The field of image-guided surgery is rapidly changing, so at the time of writing this manuscript, the use of a PGC was valid for the indications mentioned earlier; however, some of these indications may already have been replaced by new techniques in specialised centres. In the following paragraph, the results of intraoperative gamma imaging with a PGC are presented reflecting its value in oncological practice.

7.1.4 Intraoperative Imaging Protocol

While the broad aspects of the SN procedures are similar for most indications, details may vary per indication and institution. There are two different approaches to radioguided surgery with PGC:

1. Pre-surgery lymphoscintigraphy is used to place surface marks after which the PGC is used to pinpoint the exact location in the surgical field and record the entire procedure.

2. The entire procedure is performed with the PGC; thus conventional imaging is replaced by PGC-imaging before and during surgery.

Though not performed in all institutes, pre-surgery lymphoscintigraphy with conventional cameras has a clear benefit especially in the more elaborate drainage patterns, such as in melanoma and head and neck cancer (Giammarile et al. 2013). Subsequently, the intraoperative detection is performed by positioning the PGC close to the predefined area of interest. For the majority of image acquisitions, a 60 s acquisition time is sufficient for an adequate image. This is especially the case in SN procedures where the target to background ratio is relatively high. The system that is used in our hospital, the Sentinella 102, enables us to project the centre of the PGC screen (i.e. central FOV) on the skin or in the surgical cavity with two intersecting laser lines. This feature allows for quick identification of the hotspots within the surgical field. At this moment, a new feature is being developed that superimposes the count data onto a real-time optical image of the surgical field, thus providing direct and intuitive translation of the imaging data into the surgical field. For SN procedures it is useful to perform additional image acquisitions after removal of each node to validate correct excision. The final post-excision image can serve as documentation to ensure correct excision of all SN.

7.2 OVERALL RESULTS OF INTRAOPERATIVE GAMMA IMAGING

In this section, we provide per indication a brief introduction including a short description of the procedure followed by a summary of our experiences together with the accompanying experiences from other centres. Still, it has to be noted that several different approaches are valid, and that in many instances superiority of one approach over another has not been established. All studies with patients from the Netherlands Cancer Institute are tabulated to obtain a complete overview (Table 7.1).

7.2.1 Sentinel Node Breast Cancer

Currently, breast cancer is the most frequent cancer in women with nearly 1.7 million newly diagnosed cases in 2012. As in other tumours, adequate staging is required for therapy planning, and the SN procedure is regarded as standard in this respect in early-stage breast cancer. SN are defined as the first lymph nodes to receive detached tumour cells by direct lymph drainage with the potential to grow into a metastasis. In breast cancer, for intratumoural tracer administrations, these nodes are located in the axilla (~86% of the cases) and/or the parasternal region (~14% of the cases) (Leidenius et al. 2006). Morton et al. described the first use of SN biopsies in melanoma as a less invasive technique compared to an axillary clearance and has been used now for two decades in early stage breast cancer. In an experienced multidisciplinary team, intraoperative SN identification rates of over 95% have been reported regularly (Morton et al. 1992, Giammarile et al. 2013).

In 2014, around 350 SN procedures were performed in our centre out of 600 patients treated for breast cancer. Most of these procedures were

TABLE 7.1   An Overview of All PGC Use in the Netherlands Cancer Institute for Intraoperative Purposes

Publication	Indication(s)	Patients	Study Aim
Vermeeren et al. 2009b	Urological SN procedures	20	PGC for urological SN procedures
Vermeeren et al. 2010a	Urological SN procedures	18	SPECT/CT and PGC for urological SN surgery
Vermeeren et al. 2010d	Head and neck cancer SN	25	PGC for head and neck SN surgery
Vermeeren et al. 2010c	Prostate cancer SN	50	Optimising the particle concentration
Vermeeren et al. 2010b	Prostate cancer SN	10	SN drainage patterns in recurrent cancer
Bex et al. 2010	Renal cell cancer SN	8	Feasibility of SN procedure
Vidal-Sicart et al. 2010	Preoperative breast cancer SN	52	Reproducibility lympho-scintigraphy with a PGC
Brouwer et al. 2011	Testicular cancer SN	10	SPECT/CT and PGC for testicular SN surgery
Vermeeren et al. 2011	Prostate cancer SN	55	PGC for prostate SN surgery
Brouwer et al. 2012a	Melanoma SN	11	Evaluation of a hybrid tracer
Van den Berg et al. 2012	Oral cavity cancer SN	14	Evaluation of a hybrid tracer
Brouwer et al. 2012b	Melanoma and penile cancer SN	25	Evaluation of a hybrid tracer
Vidal-Sicart et al. 2013	Melanoma SN	1	PGC for head and neck SN surgery
Frontado et al. 2013	Melanoma, penile and oral cavity SN	20	Evaluation of a hybrid tracer
Matheron et al. 2013	Vulvar cancer SN	15	Evaluation of a hybrid tracer
Brouwer et al. 2014	Penile cancer SN	65	Hybrid tracer as a replacement for blue dye
Borbon-Arce et al. 2014	Head and neck cancer SN	25	Evaluation of PGC and hybrid tracer
Van den Berg et al. 2014	Melanoma SN	104	Evaluation of a hybrid tracer
Hellingman et al. 2015	Head and neck melanoma SN	3	PGC for near injection site SN

Notes: All of these studies are performed with the Sentinella 102; PGC, portable gamma camera; SN, sentinel node.

performed using ^{99m}Tc-albumin nanocolloid only, and in certain cases, in conjunction with the traditional blue dye injection. The use of blue dye depends on the surgeon's preference and on the expected detectability of SN with the gamma probe based on lymphoscintigraphy. In 2010, a multicentre study with 52 breast cancer patients scheduled for an SN biopsy was performed. This study compared the visibility of nodes with a PGC (Sentinella 102) and a conventional gamma camera. The PGC images with a 20 cm × 20 cm FOV were acquired directly after the 2 h post-injection static images on the large FOV camera. When lead shielding of the injection site was applied (in only 43 patients), 88% of the patients had visualisation of SN using the PGC compared to 95% with the conventional large FOV gamma camera, suggesting that the detection rate with the conventional camera is slightly better (Vidal-Sicart et al. 2011). So far, no additional intraoperative studies using a PGC for breast cancer SN procedures have been performed in our centre, because surgeons feel confident and achieve high success rates in locating SN with a conventional gamma probe when good preoperative lymphoscintigraphic images are available.

Other centres have reported more extensive data on the use of PGC for SN mapping and biopsy in breast cancer. For example, Mathelin et al. described a PGC, the CarollReS (Hitachi Chemical Co. Ltd., Japan), for SN detection and depth estimation of these nodes in breast cancer SN procedures. In the 11 patients included, the PGC visualised slightly more SN than conventional lymphoscintigraphy (respectively, 16SN and 14SN), with the benefit that it could also estimate the depth of the SN based on the full width at half maximum of the signal with a strong correlation to the measured depth by the surgeon (Mathelin et al. 2007). PGC are also used for primary breast tumour localisation in radioguided occult lesion localisation, where ^{99m}Tc-microaggregated albumin is injected central in the tumour, while using the Sentinella 102 or the TReCam (Bobigny University, Bondi, France; Paredes et al. 2008, Bricou et al. 2015).

7.2.2 Sentinel Node Cutaneous Melanoma

SN melanoma is, together with breast cancer, the most prevalent indication for SN procedure, as a negative SN biopsy is the most important prognostic factor for disease-free survival in stage I–II melanoma (van der Ploeg et al. 2010). The SN procedure is generally advised in clinically localised invasive melanoma (T1b–T4b, N0 and M0) (Chakera et al. 2009).

This is also the case in our centre where SN procedures for melanoma are common practice for tumour staging and optimal selection of patients before complete regional node dissection. In all cases, the distribution pattern of the SN is of course related to the site of the primary tumour and can become quite complex with visualisation of multiple SN and higher echelon nodes in several basins. In our experience, which is also underlined by other institutes, SPECT/CT imaging prior to SN biopsy is vital in difficult drainage patterns, like in the case of melanoma and head and neck cancers (Valdés Olmos et al. 2014a).

In three studies performed in our institute, including a case study, the use of a PGC for SN melanoma procedures was reported. In a first study, in 3 out of 16 patients (10 melanoma head and neck area, 6 melanoma trunk) additional SNs were localised with the PGC compared to the preoperative SPECT/CT scan (Brouwer et al. 2012a). In a second cohort of 104 patients with melanoma in the head and neck, on the trunk or extremities, the PGC was implemented as standard care in addition to preoperative lymphoscintigraphy and SPECT/CT.

The primary aim of this study was evaluation of multimodal surgical guidance using the hybrid tracer and fluorescent imaging. The clinical results of the PGC as a stand-alone device were not discussed, as it was not the research question of this study (van den Berg et al. 2014).

7.2.3 Sentinel Node Head and Neck Cancer

SN mapping for head and neck cancer in case of melanoma and Merkel cell cancer is widely used since the introduction of SN biopsies. For oral and oropharyngeal cavity cancers, predominantly squamous cell carcinoma accounting for more than 274,000 new cases annually, SN biopsies are also increasingly used as a diagnostic or staging tool (Alkureishi et al. 2009, 2010, Chakera et al. 2009, Heuveling et al. 2015). SN biopsies are indicated for clinical and radiological node negative (N0) patients with localised disease and its indication further depends on the Breslow thickness for melanoma. Nowadays, mainly T1–T2 tumours are included, since patients with these tumour stages are at high risk for lymph node metastases (Paleri et al. 2005, van der Ploeg et al. 2010).

Though, the SN procedure is technically feasible in head and neck cancers with a gamma probe, it has encountered several technical challenges in terms of unpredictable drainage patterns, near injection site SN and a highly complex and fragile anatomy within this region (Hellingman et al. 2015). These challenges led to the implementation of a PGC to aid these procedures in our centre.

Our first study using a PGC in SN procedures for head and neck cancer was published in 2010 by Vermeeren et al. (2010d). In this study, the Sentinella was used in 25 patients with either a melanoma or oral cavity carcinoma during surgery. All SN identified by means of planar lymphoscintigraphy or SPECT/CT were also localised with the PGC. After excision of the SN, the PGC was used to determine the distribution of the remaining radioactivity. In six patients, the SN was identified more efficiently in terms of localisation at complex sites by the PGC, and in nine patients, additional SNs were detected by the PGC, of which one SN was tumour positive at pathological examination.

In another study by Borbón-Arce et al., 25 patients were evaluated by a multimodality approach in which planar, SPECT/CT and PCG imaging were used. A total of 67 SN were visualised on preoperative imaging. Intraoperatively, all of these 67 SN were removed together with 22 additional nodes; 12 were located in the vicinity of the injection site during the excision, and 10 SN were located by post-excision PGC imaging (Borbón-Arce et al. 2014). In a case series, it was reported that these so-called near-injection-site SN could be detected using close-up imaging with a PGC, though not located with conventional planar and SPECT/CT imaging (Hellingman et al. 2015). A similar conclusion was also described in a case report of a patient with a melanoma located on the cutaneous preauricular area with preoperative non-visualisation on both planar imaging and SPECT/CT where the PGC did visualise an SN in close proximity to the injection site (Vidal-Sicart et al. 2013). The additionally localised SN can be of additional clinical value when they are the only positive nodes for the resected SN.

Nevertheless, we are not the only institute using this method for these procedures. In 2005, Tanaka et al. published a case report with the first image of a laryngeal cancer, and Tsuchimochi et al. published in 2008 a series of eight patients for head and neck SN procedures, both using a different PGCs than the PGC used in our centre (Tanaka et al. 2005, Tsuchimochi et al. 2008). In all of the nine patients, all indicated SN by the conventional gamma camera were detected with the PGC (Figure 7.4).

7.2.4 Sentinel Node Penile Cancer

The management of the regional lymph nodes is dependent on tumour-stage in penile cancer, so SN mapping and biopsy is generally performed in patients with clinically normal inguinal nodes (cN0) (Hakenberg et al. 2015).

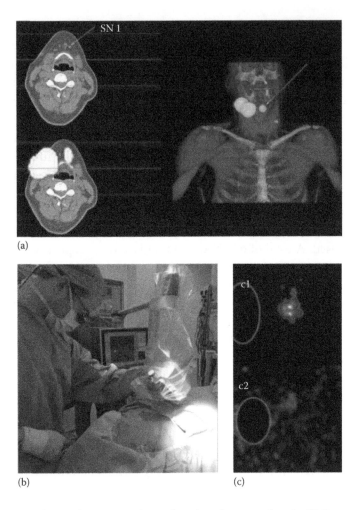

FIGURE 7.4 **(See colour insert.)** Head and neck sentinel node (SN) procedure. A 54-year-old patient with a melanoma in the right side of the neck. (a) 3D volume rendering of the SPECT/CT scan. The injection site is clearly visible and an SN at level Ia and a contralateral SN in level IV left. (b) Intraoperative imaging with a PGC, the PGC is aimed at the SN in level Ia. (c) Pre- and post-excision images with the PGC. The blackout zone used in image c1 and c2 is to shield the remaining radiation from the location of the injection site.

The SN prone to metastasis are generally located at the superficial and deep inguinal node basins. In this tumour type, evaluation of both sites of the groin is performed as non-visualisation of the SN at one or both sites can result at our institute in either uni- or bilateral complete lymph node dissection (if localisation with blue dye is unsuccessful as well). Only 20%–25% of the men with clinically normal nodes have regional metastasis, and therefore, a complete lymph node dissection may be overtreatment in the majority with considerable morbidity (Horenblas et al. 2000). Accordingly, this can be avoided by an adequate SN mapping and biopsy procedure.

An SN with the PGC or gamma probe for early stage penile cancer is currently the standard practice in our centre. In two studies described by Brouwer et al., in 2012 (9 patients) and 2014 (65 patients), a PGC was used for SN identification in penile cancer. In the first study, 10 additional SNs were located in 9 patients using a PGC compared to conventional detection methods including SPECT/CT (Brouwer et al. 2012b). In a second study, 2 years later the PGC was used to acquire pre- and post-excision images. In 22 of the 65 patients, additional nodes were depicted using the PGC compared to conventional techniques and, after the surgical area was explored once again, an additional 37 nodes were identified (Brouwer et al. 2014). Consequently, a PGC is nowadays routinely used in clinical practice for all penile cancer SN procedures at our institute.

7.2.5 Sentinel Node Vulvar Cancer

As in penile cancer, a thorough evaluation of lymph nodes in the groins for optimal staging in vulvar cancer can be considered an alternative for a complete inguinofemoral lymphadenectomy in locally advanced disease. Already in 1994, Levenback et al. introduced the SN procedure for vulvar cancer (Levenback et al. 2001). At present, the SN procedure for this indication is still used and has been reviewed as an accurate method for staging clinically and radiological node negative early stage vulvar cancer (Hassanzade et al. 2013). A systematic review by Selman et al. (2005) compared various approaches for detecting lymph node metastasis in women with vulvar cancer and concluded that the radiotracer procedure has a sensitivity of around 95%. Therefore, we incorporated this SN procedure in our clinic, and we have performed SN procedures in almost 100 patients since 2012 for this indication.

In 2013, Mathéron et al. published a study on 15 patients with vulvar cancer referred for SN detection using a PGC. PGC guided the excision location and it was used for post-excision validation of complete excision (Mathéron et al. 2013). In contrast with penile cancer and based on the experiences of the surgeon, the PGC was not used routinely after this study for all vulvar cancer SN procedures, but only in specific cases such as poor visualisation or unclear drainage during the SN procedure. In other cases, standard probes are used with great success.

7.2.6 Sentinel Node Prostate Cancer

Although validated, but not widely used, the SN procedure for prostate cancer can be an accurate method for staging patients with a Gleason score of up to 8 (Holl et al. 2009). The main reason for the limited routine use of SN biopsy in prostate cancer is probably related to the large variability in lymphatic drainage patterns. Also, there seems to be variation in the definitions of pelvic drainage basins in which metastatic SN can be found between institutions. In our institute, the drainage pattern of an individual patient on preoperative imaging is leading and can extend up to para-aortic SN nodes (Meinhardt 2007). The surgical approach varies in terms of open and closed surgery by means of laparoscopy. Currently, we perform the SN procedure for prostate cancer on a regular basis by means of laparoscopic surgery.

In different studies from our institute published in 2009, 2010 and 2011, the use of a PCG was described in this laparoscopic setting in 16, 8, 50 and 10 patients, respectively. Imaging with the PGC was performed transabdominal and the position of the laparoscopic gamma probe tip on the screen was determined by stitching an ^{125}I-seed at the tip of the gamma probe, which was visualised simultaneously on the PGC screen. The radioactive seed is imaged simultaneously with ^{99m}Tc and clearly visible on the screen of the PGC. ^{99m}Tc is visualised as a static image and the location of the ^{125}I is repeatedly updated and visualised by a moving circle as an indicator of its location on the screen. The distance between the gamma probe tip and the ^{99m}Tc hotspot can be determined in 2D (to our knowledge this feature is only present at the Sentinella 102). The aim of the different studies was respectively to determine the feasibility of radioguidance during surgery, examine the para-aortic drainage patterns, optimise the colloid particle concentration and compare treated with untreated prostate cancer patients with regard to the drainage pattern and locations (Vermeeren et al. 2009b, 2010a–c; Figure 7.5).

In a large study, including 55 patients, the additional value of a PGC for intraoperative SN visualisation was analysed. In this study, the PGC initially visualised 16% less SN compared to SPECT/CT, but during surgery, 17 additional SNs were visualised with the PGC after excision monitoring of the first SN. Two of the additionally excised SN were tumour-positive lymph nodes; nonetheless, these two patients had already positive nodes in the conventionally removed SN (Vermeeren et al. 2011). The main reason for non-visualisation with the PGC in the laparoscopic setting was the

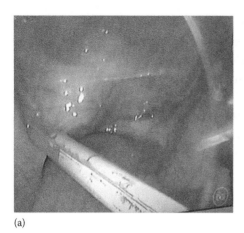

(a) (b)

FIGURE 7.5   Prostate sentinel node (SN) procedure. (a) Laparoscopic gamma probe in a laparoscopic setting. (b) Portable gamma camera overview image with the injection site at the bottom of the screen and 3 SN. The gamma probe tip position is indicated with the dotted circle projected on top of the para-aortic SN.

relatively low activity in the SN together with the relatively large imaging distance (Vermeeren et al. 2010c).

In 2012, a larger series of 121 patients underlined the relevance of SN locations outside the extended dissection area in patients who opt for external beam radiotherapy. In this study, the PGC was used for all patients although the description of the effectiveness of a PGC was limited, but it nicely demonstrates the level of clinical adaptation and the frequency of use for this method (Meinhardt et al. 2011). Of the entire population, 31% had SN outside the standard extended dissection area, however solitary metastatic nodes within this area were rare. Still, when opting for salvage therapy, laparoscopic SN procedures are feasible.

7.2.7 Sentinel Node Renal Cell Cancer

SN procedures for renal cell cancer are controversial and the survival benefit for these patients is still unclear. In addition, its application in this tumour type is also known to harbour a highly unpredictable lymphatic drainage pattern (Bex et al. 2010, Karmali et al. 2014). Nevertheless, we do occasionally perform this procedure in T1–T2 N0 patients for study purposes.

In total, 14 patients with renal cell cancer were operated on using a PGC for SN detection in renal cell cancer. One patient underwent laparoscopic surgery; thus, the setup was similar in one patient to the previously described method where an ^{125}I seed was fixed on the laparoscopic gamma probe for probe navigation. The other cases were performed during open surgery. The primary aim of the study was to visualise drainage patterns and it was found that the SN location was mainly in the para- and inter-aortocaval region. Furthermore, the feasibility of performing SN procedures for this type of procedures was demonstrated, but additional studies for this indication have not been performed as yet (Vermeeren et al. 2009b, 2010a, Bex et al. 2010).

7.2.8 Sentinel Node Testicular Cancer

Like renal cell cancer SN procedures, testicular cancer SN procedures remain a matter of discussion. However, it is unquestionable that diagnostic techniques are needed to assess N-stage at an early stage preventing unnecessary treatment in those without dissemination. Therefore, the SN procedure was introduced for stage I patients with testicular cancer in our hospital (Tanis et al. 2002, Brouwer et al. 2011). Because of its low incidence, in only nine patients undergoing surgery for testicular cancer, a PGC for post-excision validation was used. In 20% of these patients, additional nodes were localised by using this PGC. These results were described in three studies including, respectively, two patients, four patients, whereas Brouwer et al. described ultimately nine patients in 2011 including also the six patients from the previous studies (Vermeeren et al. 2009b, 2010a, Brouwer et al. 2011). Testicular SN procedures as well as renal cell cancer SN procedures are relatively rare in our hospital, and therefore, a PGC can be especially helpful to ensure complete excisions and support the surgeon in making decisions.

7.2.9 Isolated Limb Perfusion Scans

Local treatment with high-dose chemotherapy, often tumour necrosis factor (TNF-alpha) combined with interferon gamma and melphalan, of malignant melanoma or sarcoma of the limb using extremity perfusion is an elegant and effective treatment. At our institute, roughly 30 isolated limb perfusion treatments are performed as a standard clinical procedure in which the PGC is used to monitor the procedure. In 2009, Orero et al. published a complete description of the procedure with a PGC and its clinical use for this indication (Orero et al. 2009). In short, the major arteries and veins of the limb are clamped and connected to an oxygenated extracorporeal perfusion circuit; the smaller superficial vessels are clamped with a tight tourniquet. High doses of chemotherapeutical are circulated within the limb, so any leakage from this isolated blood territory can result in a high degree of systemic toxicity (Grünhagen et al. 2006). In this setup, a radioactive tracer, often radiolabelled serum albumin or erythrocytes, is circulated in the isolated extremity and the systemic circulating radioactivity concentration is monitored with a PGC positioned above the heart. The PGC is equipped with a specially designed flat-field collimator to maximise the incoming counts, while limiting the effects of scatter from the limb. In case of a slowly increasing radioactivity concentration in the systemic blood pool an additional vessel restriction can be applied or, when the increase is too fast, the procedure should be stopped in order to avoid any systemic toxicity (Orero et al. 2009).

7.3 RESULTS COMPARED TO OTHER TECHNIQUES

Other competitive or complementary techniques do exist and facilitate the same or a similar purpose. For example, freehand-SPECT (declipseSPECT, SurgicEye GbmH, Munich, Germany) is another type of intraoperative imaging that provides additional depth indication by a 3D reconstruction of the radioactive target lesions (see Chapter 4 for more information). Currently, different studies are performed to determine the optimal intraoperative imaging method with one of the two or with both techniques (Bluemel et al. 2013, Casáns-Tormo et al. 2015).

A different, non-radioactive, approach is fluorescence-guided imaging, which has a very high resolution for close-up imaging using fluorescent cameras, but this method lacks the possibility of preoperative imaging. To overcome this problem, a hybrid fluorescent-radioactive tracer (indocyanine green ^{99m}Tc-nanocolloid) has been introduced in our institute. This hybrid tracer is both fluorescent and radioactive and therefore allows preoperative SN imaging, real-time radioguidance and close-up high-resolution fluorescent guidance (Valdés Olmos et al. 2014a). In two studies at our institute, this hybrid tracer was used in head and neck surgery together with a PGC to provide excision conformance through pre- and post-excision images. The hybrid tracer demonstrated similar drainage patterns as the standard radioactive tracer (^{99m}Tc-nanocolloid), but also demonstrated its additional value for detecting SN situated close to injection site (Brouwer et al. 2012a, van den Berg et al. 2012, Frontado et al. 2013). Other groups have also successfully implemented this approach for head and neck malignancies.

Another important clinical application of PGC, which is not mentioned in this chapter, is its use for thyroid and parathyroid surgery; these procedures are not performed in our institute, but other institutes have shown excellent results (Ortega et al. 2007, Estrems et al. 2012).

7.4 DISCUSSION AND FUTURE APPLICATIONS

Following from our previous publications, the integration of PGC for complex radioguided procedures is strongly encouraged for adequate surgery. It provides pre-, peri- and postoperative guidance by SFOV images of the targets. A concept was presented to aid a successful introduction of novel techniques used for radioguided surgery. In 2011, the concept of guided intraoperative scintigraphic tumour targeting was delineated by experts in this field to provide a roadmap that would optimise surgery by radioguidance. It was advised to aim for close collaborations between institutes by forming mixed networks consisting of universities and clinical and commercial partners. In our experience, this has been a valuable aspect in the introduction of radioguided surgery in daily clinical work (Zaknun et al. 2012, Valdés Olmos et al. 2014b, Vidal-Sicart et al. 2014).

In addition to our own experience, there is an increasing number of other expertise centres using or implementing PGCs for various indications. It has to be stressed that all our recent clinical results are based on just one system, the Sentinella 102. We only have limited clinical experiences with comparable systems described in technical and clinical analyses by other centres, and other systems might demonstrate different clinical results due to differences in specifications (Sánchez et al. 2006, Bugby et al. 2014). It is important for new users to evaluate the needs in terms of FOV, sensitivity, image resolution and financial plan before deciding which system would be most appropriate. We would strongly encourage a site-visit to a centre already using PGCs in clinical practice, to view performance in daily practice before making the final decision.

An optional feature is now available as upgrade to the Sentinella 102 enabling another hybrid approach. Optical imaging is added to the scintigraphic image for anatomical orientation (Hellingman et al. 2016). In addition, a novel camera characterised by Bugby et al. uses the same hybrid approach with an integrated optical module for optical and scintigraphic imaging (Lees et al. 2011, Bugby et al. 2013, 2014). The hybrid approach may be beneficial for image interpretation and anatomical localisation of certain targets, but this is not as yet established (Figure 7.6).

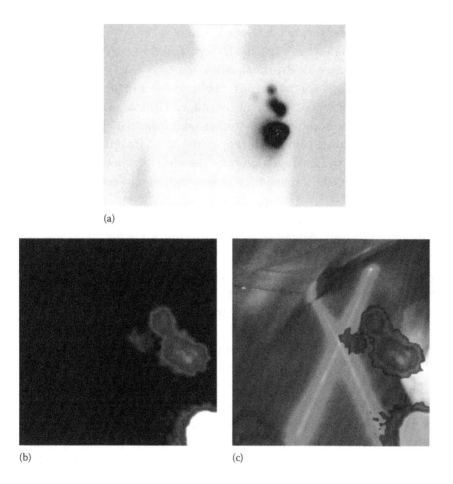

(a)

(b)   (c)

FIGURE 7.6 **(See colour insert.)** (a) Conventional lymphoscintigraphy, an injection site with a cluster of sentinel node cranial in this image located in the axilla. (b) Scintigraphic image acquired with a portable gamma camera. (c) Example of a hybrid image. In this image the scintigraphic image and the optical image are displayed in a single view.

For future applications, we explore the possibility of replacing preoperative imaging entirely at the department of nuclear medicine by PGC imaging. Certain standardised indications, like SN procedures for primary breast cancer, could be managed using only SFOV imaging by PGC. For now, this is still in the research phase at our institute but theoretically this could drastically enhance imaging logistics before surgery. Eventually, the complete SN procedure might even (partially) be replaced by a radioguided intervention in an outpatient setting. Imaging modalities, when adequately combined and developed, could accurately provide needle biopsy navigation towards specific targets, and thereby, circumvent surgery in specific cases (de Bree et al. 2015).

Although there are numerous publications and many encouraging developments, PGC use is not standard clinical practice in many centres. The lack of multicentre studies with large patient cohorts that prove the additional value of intraoperative PGC imaging could be the reason for the current adaptation rate. There is still a need for evidence-based studies before the majority of hospitals would consider such an investment.

A large systemic review by Ahmed et al. published in *Lancet Oncology* not only describes novel techniques for breast cancer SN procedures, but also points out an important obstacle for radioguided surgery; it has to be realised that currently only 60% of the eligible people have access to radioisotopes because of the complex logistics and legislation required for handling radioactive substances for medical use (Ahmed et al. 2014). Therefore, novel SN localisation methods without radioactivity have been presented. SN biopsies based on microbubbles localised with ultrasonography (Sever et al. 2009), super paramagnetic iron oxide localised using a handheld magnetometer (Douek et al. 2014) and near-infrared fluorescence imaging using indocyanine green and a fluorescence camera (van der Vorst et al. 2012) have been reported and demonstrated promising results for non-radioactive intraoperative SN localisation (Ahmed et al. 2014). Disadvantages of these techniques are the lack of preoperative imaging for SN mapping; lymphoscintigraphy is able to discriminate between first echelon and second echelon nodes. MR imaging while using iron particles might solve this imaging problem but for now this would have great influence on the workload of the already frequently used MR systems. Intraoperative SFOV imaging methods for these tracers similar to PGC imaging would be a great supplement for these techniques; however, this is currently unavailable.

In our centre, we observe a shift from non-specific tracers towards more specific tracers. An example is radiolabelled prostate specific membrane antigen. In this way imaging and radioguided surgery are enabled towards specific target lesions. We expect that PGCs could play an important role in this development by means of close-up high-resolution imaging of specific lesions during surgery or interventions and preliminary studies are initiated to evaluate this concept.

7.5 CONCLUSION

PGCs are known for their better spatial resolution (i.e. resolving power or ability to distinguish two hot spots) compared to a gamma probe and provide a real-time overview of the radioactivity distribution in the imaged area. An additional benefit of surgery with a PGC is the ability to record all the different steps in a procedure. The contribution of PGCs for clinical procedures is common use in our institute. For certain surgical procedures the addition of a PGC led to improved SN detection and especially aided SN localisation in complex areas such as in head and neck surgery and near injection site SN locations. It is expected that PGC use will increase in the future and hybrid and more targeted solutions will improve radioguided surgery even further.

ACKNOWLEDGEMENTS

We thank the disciplines involved in radioguided surgery at the Netherlands Cancer Institute and particularly Dr. Renato Valdés Olmos for his continuous efforts to implement this important area of clinical care and investigation.

REFERENCES

Abe, A., N. Takahashi, J. Lee et al. 2003. Performance evaluation of a hand-held, semiconductor (CdZnTe)-based gamma camera. *European Journal of Nuclear Medicine and Molecular Imaging*, 30(6):805–811.

Ahmed, M., A.D. Purushotham and M. Douek. 2014. Novel techniques for sentinel lymph node biopsy in breast cancer: A systematic review. *The Lancet Oncology*, 15(8):e351–e362.

Alkureishi, L.W., Z. Burak, J.A. Alvarez et al. 2009. Joint practice guidelines for radionuclide lymphoscintigraphy for sentinel node localization in oral/oropharyngeal squamous cell carcinoma. *Annals of Surgical Oncology*, 16(11):3190–3210.

Alkureishi, L.W., G.L. Ross, T. Shoaib et al. 2010. Sentinel node biopsy in head and neck squamous cell cancer: 5-year follow-up of a European multicenter trial. *Annals of Surgical Oncology*, 17(9):2459–2464.

Barranger, E., S. Uzan, S. Pitre, M.A. Duval and Y. Charon. 2008. Development of a hand-held gamma camera (POCI) in the sentinel node biopsy for breast cancer. *Pathologie-Biologie*, 56(5):254–256.

Bex, A., L. Vermeeren, G. de Windt, W. Prevoo, S. Horenblas and R.A. Valdés Olmos. 2010. Feasibility of sentinel node detection in renal cell carcinoma: A pilot study. *European Journal of Nuclear Medicine and Molecular Imaging*, 37(6):1117–1123.

Bluemel, C., A. Schnelzer, A. Okur et al. 2013. Freehand SPECT for image-guided sentinel lymph node biopsy in breast cancer. *European Journal of Nuclear Medicine and Molecular Imaging*, 40(11):1656–1661.

Borbón-Arce, M., O.R. Brouwer, N.S. van den Berg et al. 2014. An innovative multimodality approach for sentinel node mapping and biopsy in head and neck malignancies. *Revista Española de Medicina Nuclear e Imagen Molecular (English Edition)*, 33(5):274–279.

Bricou, A., M.A. Duval, L. Bardet et al. 2015. Is there a role for a handheld gamma camera (TReCam) in the SNOLL breast cancer procedure? *Quarterly Journal of Nuclear Medicine and Molecular Imaging*. Epub ahead of print.

Brouwer, O.R., T. Buckle, L. Vermeeren et al. 2012b. Comparing the hybrid fluorescent–radioactive tracer indocyanine green–^{99m}Tc-nanocolloid with ^{99m}Tc-nanocolloid for sentinel node identification: A validation study using lymphoscintigraphy and SPECT/CT. *Journal of Nuclear Medicine*, 53(7):1034–1040.

Brouwer, O.R., W.M.C. Klop, T. Buckle et al. 2012a. Feasibility of sentinel node biopsy in head and neck melanoma using a hybrid radioactive and fluorescent tracer. *Annals of Surgical Oncology*, 19(6):1988–1994.

Brouwer, O.R., R.A. Valdés Olmos, L. Vermeeren, C.A. Hoefnagel, O.E. Nieweg and S. Horenblas. 2011. SPECT/CT and a portable γ-camera for image-guided laparoscopic sentinel node biopsy in testicular cancer. *Journal of Nuclear Medicine*, 52(4):551–554.

Brouwer, O.R., N.S. van den Berg, H.M. Mathéron et al. 2014. A hybrid radioactive and fluorescent tracer for sentinel node biopsy in penile carcinoma as a potential replacement for blue dye. *European Urology*, 65(3):600–609.

Bugby, S., J. Lees, D. Bassford, E. Blackshaw and A. Perkins. 2013. Preliminary evaluation of a novel high-resolution hybrid gamma/optical camera for intraoperative imaging. In *European Journal of Nuclear Medicine and Molecular Imaging*, Vol. 40, pp. S177–S177, New York.

Bugby, S.L., J.E. Lees, B.S. Bhatia and A.C. Perkins. 2014. Characterisation of a high resolution small field of view portable gamma camera. *Physica Medica*, 30(3):331–339.

Casáns-Tormo, I., S. Prado-Wohlwend, R. Díaz-Expósito, N. Cassinello-Fernández and J. Ortega-Serrano. 2015. Initial experience in the intra-operative radiolocalization of the parathyroid adenoma with freehand SPECT and comparative assessment with portable gamma-camera. *Revista Española de Medicina Nuclear e Imagen Molecular (English Edition)*, 34(2):116–119.

Chakera, A.H., B. Hesse, Z. Burak et al. 2009. EANM-EORTC general recommendations for sentinel node diagnostics in melanoma. *European Journal of Nuclear Medicine and Molecular Imaging*, 36(10):1713–1742.

de Bree, R., B. Pouw, D.A. Heuveling and J.A. Castelijns. 2015. Fusion of freehand SPECT and ultrasound to perform ultrasound-guided fine-needle aspiration cytology of sentinel nodes in head and neck cancer. *American Journal of Neuroradiology*, 36(11):2153–2158.

Douek, M., J. Klaase, I. Monypenny et al. 2014. Sentinel node biopsy using a magnetic tracer versus standard technique: The SentiMAG multicentre trial. *Annals of Surgical Oncology*, 21(4):1237–1245.

Estrems, P., F. Guallart, P. Abreu, P. Sopena, J. Dalmau and R. Sopena. 2012. The intraoperative mini gamma camera in primary hyperparathyroidism surgery. *Acta Otorrinolaringologica (English Edition)*, *63*(6):450–457.

Frontado, L.M., O.R. Brouwer, N.S van den Berg, H.M. Mathéron, S. Vidal-Sicart, F.W.B. van Leeuwen and R.A. Valdés Olmos. 2013. Added value of the hybrid tracer indocyanine green-^{99m}Tc-nanocolloid for sentinel node biopsy in a series of patients with different lymphatic drainage patterns. *Revista Española de Medicina Nuclear e Imagen Molecular (English Edition)*, *32*(4):227–233.

Giammarile, F., N. Alazraki, J.N. Aarsvold et al. 2013. The EANM and SNMMI practice guideline for lymphoscintigraphy and sentinel node localization in breast cancer. *European Journal of Nuclear Medicine and Molecular Imaging*, *40*(12):1932–1947.

Grünhagen, D.J., J.H. de Wilt, T.L. ten Hagen and A.M. Eggermont. 2006. Technology insight: Utility of TNF-α-based isolated limb perfusion to avoid amputation of irresectable tumors of the extremities. *Nature Clinical Practice Oncology*, *3*(2):94–103.

Hakenberg, O.W., E.M. Compérat, S. Minhas, A. Necchi, C. Protzel and N. Watkin. 2015. EAU guidelines on penile cancer: 2014 update. *European Urology*, *67*(1):142–150.

Hassanzade, M., M. Attaran, G. Treglia, Z. Yousefi and R. Sadeghi. 2013. Lymphatic mapping and sentinel node biopsy in squamous cell carcinoma of the vulva: Systematic review and meta-analysis of the literature. *Gynecologic Oncology*, *130*(1):237–245.

Hellingman, D., L.J. de Wit-van der Veen, W.M.C. Klop and R.A. Valdés Olmos. 2015. Detecting near-the-injection-site sentinel nodes in head and neck melanomas with a high-resolution portable gamma camera. *Clinical Nuclear Medicine*, *40*(1):e11–e16.

Hellingman, D., S. Vidal-Sicart, P. Paredes and R.A. Valdés Olmos. 2016. A new portable hybrid camera for fused optical and scintigraphic imaging: First clinical experiences. *Clinical Nuclear Medicine*, *41*(1):e39–e43.

Heuveling, D.A., S. van Weert, K.H. Karagozoglu and R. de Bree. 2015. Evaluation of the use of freehand SPECT for sentinel node biopsy in early stage oral carcinoma. *Oral Oncology*, *51*(3):287–290.

Holl, G., R. Dorn, H. Wengenmair, D. Weckermann and J. Sciuk. 2009. Validation of sentinel lymph node dissection in prostate cancer: Experience in more than 2,000 patients. *European Journal of Nuclear Medicine and Molecular Imaging*, *36*(9):1377–1382.

Horenblas, S., L. Jansen, W. Meinhardt, C.A. Hoefnagel and O.E. Nieweg. 2000. Detection of occult metastasis in squamous cell carcinoma of the penis using a dynamic sentinel node procedure. *The Journal of Urology*, *163*(1):100–104.

Karmali, R.J., H. Suami, C.G. Wood and J.A. Karam. 2014. Lymphatic drainage in renal cell carcinoma: Back to the basics. *British Journal of Urology International*, *114*(6):806–817.

Knoll, P., S. Mirzaei, K. Schwenkenbecher and T. Barthel. 2014. Performance evaluation of a solid-state detector based handheld gamma camera system. *Frontiers in Biomedical Technologies*, 1(1):61–67.

Lees, J.E., D.J. Bassford, O.E. Blake, P.E. Blackshaw and A.C. Perkins. 2011. A high resolution Small Field Of View (SFOV) gamma camera: A columnar scintillator coated CCD imager for medical applications. *Journal of Instrumentation*, 6(12):C12033.

Leidenius, M.H.K., L.A. Krogerus, T.S. Toivonen, E.A. Leppänen and K.A.J. von Smitten. 2006. The clinical value of parasternal sentinel node biopsy in breast cancer. *Annals of Surgical Oncology*, 13(3):321–326.

Levenback, C., R.L. Coleman, T.W. Burke, D. Bodurka-Bevers, J.K. Wolf and D.M. Gershenson. 2001. Intraoperative lymphatic mapping and sentinel node identification with blue dye in patients with vulvar cancer. *Gynecologic Oncology*, 83(2):276–281.

Mathelin, C., S. Salvador, D. Huss and J.L. Guyonnet. 2007. Precise localization of sentinel lymph nodes and estimation of their depth using a prototype intraoperative mini γ-camera in patients with breast cancer. *Journal of Nuclear Medicine*, 48(4):623–629.

Mathéron, H.M., N.S. van den Berg, O.R. Brouwer et al. 2013. Multimodal surgical guidance towards the sentinel node in vulvar cancer. *Gynecologic Oncology*, 131(3):720–725.

Meinhardt, W. 2007. Sentinel node evaluation in prostate cancer. *EAU-EBU Update Series*, 5(6):223–231.

Meinhardt, W., H.G. van der Poel, R.A.Valdés Olmos, A. Bex, O.R. Brouwer and S. Horenblas. 2011. Laparoscopic sentinel lymph node biopsy for prostate cancer: The relevance of locations outside the extended dissection area. *Prostate Cancer*, 2012(4):751–753.

Morton, D.L., D.R. Wen, J.H. Wong et al. 1992. Technical details of intraoperative lymphatic mapping for early stage melanoma. *Archives of Surgery*, 127(4):392–399.

Orero, A., S. Vidal-Sicart, N. Roé et al. 2009. Monitoring system for isolated limb perfusion based on a portable gamma camera. *Nuklearmedizin*, 48(4):166–172.

Ortega, J., J. Ferrer-Rebolleda, N. Cassinello and S. Lledo. 2007. Potential role of a new hand-held miniature gamma camera in performing minimally invasive parathyroidectomy. *European Journal of Nuclear Medicine and Molecular Imaging*, 34(2):165–169.

Otake, H., T. Higuchi, Y. Takeuchi et al. 2002. Evaluation of efficiency of a semiconductor gamma camera. *Kaku igaku. The Japanese Journal of Nuclear Medicine*, 39(4):549–553.

Paleri, V., G. Rees, P. Arullendran, T. Shoaib and S. Krishman. 2005. Sentinel node biopsy in squamous cell cancer of the oral cavity and oral pharynx: A diagnostic meta-analysis. *Head & Neck*, 27(9):739–747.

Paredes, P., S. Vidal-Sicart, G. Zanón et al. 2008. Radioguided occult lesion localisation in breast cancer using an intraoperative portable gamma camera:

First results. *European Journal of Nuclear Medicine and Molecular Imaging*, 35(2):230–235.

Russo, P., A.S. Curion, G. Mettivier et al. 2011. Evaluation of a CdTe semiconductor based compact gamma camera for sentinel lymph node imaging. *Medical Physics*, 38(3):1547–1560.

Sánchez, F., M.M. Fernández, M. Giménez et al. 2006. Performance tests of two portable mini gamma cameras for medical applications. *Medical Physics*, 33(11):4210–4220.

Selman, T.J., D.M. Luesley, N. Acheson, K.S. Khan and C.H. Mann. 2005. A systematic review of the accuracy of diagnostic tests for inguinal lymph node status in vulvar cancer. *Gynecologic Oncology*, 99(1):206–214.

Sever, A., S. Jones, K. Cox, J. Weeks, P. Mills and P. Jones. 2009. Preoperative localization of sentinel lymph nodes using intradermal microbubbles and contrast-enhanced ultrasonography in patients with breast cancer. *British Journal of Surgery*, 96(11):1295–1299.

Soluri, A. and R. Pani. 2001. Miniaturized gamma camera with very high spatial resolution. U.S. Patent 6,242,744.

Tanaka, C., H. Fujii, A. Shiotani, Y. Kitagawa, K. Nakamura and A. Kubo. 2005. Sentinel node imaging of laryngeal cancer using a portable gamma camera with CdTe semiconductor detectors. *Clinical Nuclear Medicine*, 30(6):440–443.

Tanis, P.J., S. Horenblas, R.A. Valdés Olmos, C.A. Hoefnagel and O.E. Nieweg. 2002. Feasibility of sentinel node lymphoscintigraphy in stage I testicular cancer. *European Journal of Nuclear Medicine and Molecular Imaging*, 29(5):670–673.

Tsuchimochi, M., K. Hayama, T. Oda, M. Togashi and H. Sakahara. 2008. Evaluation of the efficacy of a small CdTe γ-camera for sentinel lymph node biopsy. *Journal of Nuclear Medicine*, 49(6):956–962.

Valdés Olmos, R.A., D.D. Rietbergen, S. Vidal-Sicart, G. Manca, F. Giammarile and G. Mariani. 2014a. Contribution of SPECT/CT imaging to radioguided sentinel lymph node biopsy in breast cancer, melanoma, and other solid cancers: From "open and see" to "see and open". *Quarterly Journal of Nuclear Medicine and Molecular Imaging*, 58:127–39.

Valdés Olmos, R.A., S. Vidal-Sicart, F. Giammarile, J.J. Zaknun, F.W.B. Van Leeuwen and G. Mariani. 2014b. The GOSTT concept and hybrid mixed/virtual/augmented reality environment radioguided surgery. *Quarterly Journal of Nuclear Medicine and Molecular Imaging*, 58(2):207–215.

van den Berg, N.S., O.R. Brouwer, W.M.C. Klop et al. 2012. Concomitant radio- and fluorescence-guided sentinel lymph node biopsy in squamous cell carcinoma of the oral cavity using ICG-^{99m}Tc-nanocolloid. *European Journal of Nuclear Medicine and Molecular Imaging*, 39(7):1128–1136.

van den Berg, N.S., O.R. Brouwer, B.E. Schaafsma et al. 2014. Multimodal surgical guidance during sentinel node biopsy for melanoma: Combined gamma tracing and fluorescence imaging of the sentinel node through use of the hybrid tracer indocyanine green–^{99m}Tc-nanocolloid. *Radiology*, 275(2):521–529.

van der Ploeg, A.P., A.C. van Akkooi, P.I. Schmitz, S. Koljenovic, C. Verhoef and A.M. Eggermont. 2010. EORTC Melanoma Group sentinel node protocol identifies high rate of submicrometastases according to Rotterdam Criteria. *European Journal of Cancer, 46*(13):2414–2421.

van der Vorst, J.R., B.E. Schaafsma, F.P. Verbeek et al. 2012. Randomized comparison of near-infrared fluorescence imaging using indocyanine green and ^{99m}technetium with or without patent blue for the sentinel lymph node procedure in breast cancer patients. *Annals of Surgical Oncology, 19*(13):4104–4111.

Vermeeren, L., W.M.C. Klop, M.W.M. Van den Brekel, A.J.M. Balm, O.E. Nieweg and R.A. Valdés Olmos. 2009a. Sentinel node detection in head and neck malignancies: Innovations in radioguided surgery. *Journal of Oncology, 2009*:681746.

Vermeeren, L., W. Meinhardt, A. Bex et al. 2010a. Paraaortic sentinel lymph nodes: Toward optimal detection and intraoperative localization using SPECT/CT and intraoperative real-time imaging. *Journal of Nuclear Medicine, 51*(3):376–382.

Vermeeren, L., W. Meinhardt, H.G. van der Poel and R.A. Valdés Olmos. 2010b. Lymphatic drainage from the treated versus untreated prostate: Feasibility of sentinel node biopsy in recurrent cancer. *European Journal of Nuclear Medicine and Molecular Imaging, 37*(11):2021–2026.

Vermeeren, L., S.H. Muller, W. Meinhardt and R.A. Valdés Olmos. 2010c. Optimizing the colloid particle concentration for improved preoperative and intraoperative image-guided detection of sentinel nodes in prostate cancer. *European Journal of Nuclear Medicine and Molecular Imaging, 37*(7):1328–1334.

Vermeeren, L., R.A. Valdés Olmos, W.M.C. Klop, A.J. Balm and M.W. van den Brekel. 2010d. A portable γ-camera for intraoperative detection of sentinel nodes in the head and neck region. *Journal of Nuclear Medicine, 51*(5):700–703.

Vermeeren, L., R.A. Valdés Olmos, W. Meinhardt and S. Horenblas. 2011. Intraoperative imaging for sentinel node identification in prostate carcinoma: Its use in combination with other techniques. *Journal of Nuclear Medicine, 52*(5):741–744.

Vermeeren, L., R.A. Valdés Olmos, W. Meinhardt et al. 2009b. Intraoperative radioguidance with a portable gamma camera: A novel technique for laparoscopic sentinel node localisation in urological malignancies. *European Journal of Nuclear Medicine and Molecular Imaging, 36*(7):1029–1036.

Vidal-Sicart, S., O.R. Brouwer, H.M. Mathéron, I.B. Tan and R.A. Valdés Olmos. 2013. Sentinel node identification with a portable gamma camera in a case without visualization on conventional lymphoscintigraphy and SPECT/CT. *Revista Española de Medicina Nuclear e Imagen Molecular Molecular (English Edition), 3*(32):203–204.

Vidal-Sicart, S., M.E. Rioja, P. Paredes, M.R. Keshtgar and R.A. Valdés Olmos. 2014. Contribution of perioperative imaging to radioguided surgery. *Quarterly Journal of Nuclear Medicine and Molecular Imaging, 58*:140–160.

Vidal-Sicart, S., L. Vermeeren, O. Solà, P. Paredes and R.A. Valdés Olmos. 2011. The use of a portable gamma camera for preoperative lymphatic mapping: A comparison with a conventional gamma camera. *European Journal of Nuclear Medicine and Molecular Imaging*, 38(4):636–641.

Zaknun, J.J., F. Giammarile, R.A. Valdés Olmos, S. Vidal-Sicart and G. Mariani. 2012. Changing paradigms in radioguided surgery and intraoperative imaging: The GOSTT concept. *European Journal of Nuclear Medicine and Molecular Imaging*, 39(1):1–3.

CHAPTER 8

Radioguided Surgery in Oral Cancers

Hybrid Imaging and Hybrid Tracers

C. Schilling and M. McGurk

CONTENTS

8.1 INTRODUCTION: BACKGROUND AND DRIVING FORCES

Primary oral cancer, which is predominantly (>90%) oral squamous cell carcinoma (OSCC), is the eighth most common cancer worldwide (Petersen 2003). There is considerable global variation in incidence of OSCC; in India, it is the most common cancer diagnosis in men and cases are increasing, whereas in Western countries, there is a recent small dip in incidence commensurate with a reduction in smoking (Warnakulasuriya 2009). However, there has been little improvement in overall survival in the last 50 years (Chaturvedi et al. 2013).

It is estimated that worldwide there are at least 275,000 new oral cancer diagnoses per year, up to 60% of which are considered late stage (T3-T4, N ≥ 1, M1). The presence of cervical lymph node metastasis at diagnosis is associated with a 50% reduction in cure. The salutary message is that even for early stage tumours (T1-2, N0 M0), the 5-year survival is in the region of 75%. Death from OSCC is usually from locoregional disease, and it is felt that the lower than expected survival even in early stage tumours may be due to missed or undertreated cervical metastasis (Ganly et al. 2012).

Up to two-thirds of tumours of the oral cavity are located on the anterior tongue (anterior to the circumvallate papillae including ventral and lateral surfaces) and the floor of the mouth (FOM). Lymph node metastasis that occur relatively early in the disease process predominantly but not exclusively occur to the ipsilateral neck in levels I, IIa, IIb, III and IV and rarely level V (Farmer et al. 2015). Up to 10% of patients show contralateral drainage that would be missed by conventional ipsilateral neck dissection. Preoperative staging to detect cervical metastasis may be undertaken by any combination of CT/MRI/USS ± fine-needle aspiration, but the sensitivity of these modalities in detecting small metastatic deposits is in about 70% (Stoeckli et al. 2012). For this reason, elective neck dissection (END), usually selective removal of ipsilateral neck levels I–IV, is the standard practice when there is judged to be a >20% risk of occult cervical metastasis. Neck dissection exposes the patient to a prolonged surgical procedure that can result in significant morbidity in terms of neck and shoulder function even though it is recognised that procedure is unnecessary in the majority of cases. A further complicating factor is that the pathological analysis of the large number of nodes yielded (minimum of 5 nodes per level but may be as much as 60–70 nodes in total) is inevitably limited often to a single axial section with H&E staining. This can overlook small metastatic deposits, the serious consequence of under-diagnosis being a missed opportunity for adjuvant therapy (postoperative radiotherapy or chemoradiotherapy) thereby reducing the chances of cure.

There is a clear need to focus lymph node retrieval to the 'at-risk' or sentinel nodes (SNs) allowing detailed pathological examination for accurate staging and appropriate therapy in those with metastasis, whilst sparing the majority of patients the morbidity associated with blanket nodal resection (Figure 8.1).

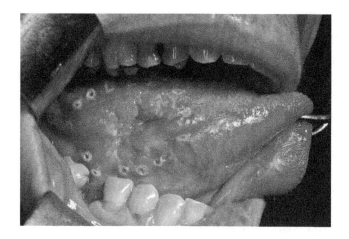

FIGURE 8.1 T1 oral squamous cell carcinoma of the right anterior tongue (central ulcer) with excision margin marked. (Photograph courtesy of M. McGurk.)

8.2 VALUE OF SENTINEL NODE BIOPSY IN ORAL CANCER

The introduction of sentinel node biopsy (SNB) into the oral cavity was developed by surgeons working in the field of head and neck surgery, who were practicing SNB routinely for melanoma. In 1999, a group in Cannisburn described their pilot technique using a combination of blue dye and radiotracer in 12 patients validated against immediate neck dissection (Shoaib et al. 1999). In this study, all seven patients with occult metastasis were correctly staged by SNB. This success led onto the first prospective multicentre trial examining the reliability of SNB which proved the technique so effective that the requirement for concurrent neck dissection was dropped during the trial (Alkureishi et al. 2010). Since this time, there has been a number of case series (Broglie et al. 2011, Melkane et al. 2012, Pezier et al. 2012, Pedersen et al. 2015) as well as multicentre studies (Burcia et al. 2010, Civantos et al. 2010, Flach et al. 2014) and meta-analysis (Paleri et al. 2005, Govers et al. 2013a) confirming the safety of SNB in oral cancer. The largest cohort of patients to undergo SNB ($n = 415$) as the sole staging tool was described in the SENT trial which showed a 99.5% SN identification rate with sensitivity of 86% and a 3-year disease specific survival of 94% (Schilling et al. 2015).

Although the use of SNB for oral cancer is becoming more common, it is far from routine practice across most countries. In Denmark, it has been incorporated into the national cancer treatment pathway, but in countries such as Germany, it must be undertaken alongside neck dissection and even then in the context of a research study. It is likely that over the next few years, there will be national recommendations promoting the use of SNB in oral cancer, but perhaps one of the reasons that there has been slow uptake is the relative paucity of data showing the benefit of SNB especially when directly compared to END.

There are a small number of studies that have looked at the value of SNB compared to END in terms of the patient morbidity and health economic profiles. Most recently, Govers et al. (2016) published data showing that SNB has significantly better quality of life and shoulder scores compared to END, confirming previous data reporting improved patient reported outcomes and significantly reduced complication profile of SNB over END (Schiefke et al. 2009, Murer et al. 2011, Hernando et al. 2014). Govers had also previously shown that SNB constituted the most cost effective treatment strategy (Govers et al. 2013b), data further supported by a study showing a cost saving if even half of the SNB proved positive for metastasis (O'Connor et al. 2013). These direct costs are modelled on the traditional SNB pathway of standard radiotracer and lymphoscintigraphy. It should therefore be born in mind that the economic advantages may be diminished by the advanced technologies that we will discuss further.

8.3 CURRENT LIMITATIONS OF SENTINEL NODE BIOPSY IN ORAL CANCER

Despite the proven reliability of SN identification in oral cancer, there has not at present been widespread adoption of the technique. One factor discouraging uptake is the high false-negative rate (FNR; calculated as the number of isolated recurrences in the neck after a negative SNB, divided by the number of true positive SNB results). In oral cancer, the FNR is variously reported between 6% and 14% using traditional radiotracers (Table 8.1). Some studies have particularly highlighted an increased FNR in relation to tumours of the FOM, and it is postulated that the close proximity of the injection site to the lymph nodes results in shine-through effect masking the SN.

The FNR in breast cancer is around 5%, although the reverse is true in melanoma where the FNR is as high as 20% (Morton et al. 2014). Nevertheless, clinicians may be unwilling to forgo END in favour of SNB even with a comparable survival and improved morbidity for SNB if there is potentially a one in ten chances of misdiagnosis.

TABLE 8.1 Comparison of Outcome of Recent Trials in Sentinel Node Biopsy in Oral Squamous Cell Carcinoma

Author (Year)	Number of Patients	Positive SNB	False-Negative Rate (Overall)	FNR for Floor of the Mouth	Overall Sensitivity %	Overall NPV %
Schilling (2015)	415	23% (94/415)	14% (15/109)	13% (2/16)	86	95
Alvarez et al. (2014)	63[a] (28 SNB)	39% (7/28)	N/A	36% (4/11))	64	81
Broglie et al. (2013)	111	38% (49/111)	6% (3/52)	Not reported	94	96
Burcia (2010)	50	36% (18/50)	0%[b] (0/18)	Not reported	100	100
Alkureishi et al. (2010)	125	34% (42/125)	8.7%[b] (4/46)	88% (4/5)	91	95
Civantos et al. (2010)	140	23.5% (33/140)	9.8%[b] (4/41)	25% (1/4)	82.5	96

[a] All patients were FOM tumours only.
[b] Some or all patients underwent concurrent neck dissection.

There is therefore a great interest in improving the accuracy of the SNB in oral cancer which has resulted in the investigation of a number of new technologies such as 3D navigation-guided SNB with freehandSEPCT (fhSPECT) (Bluemel et al. 2014), multimodal tracers incorporating both optical and gamma signal (Van Leeuwen et al. 2011) and receptor targeted tracers (Agrawal et al. 2015; Figure 8.2).

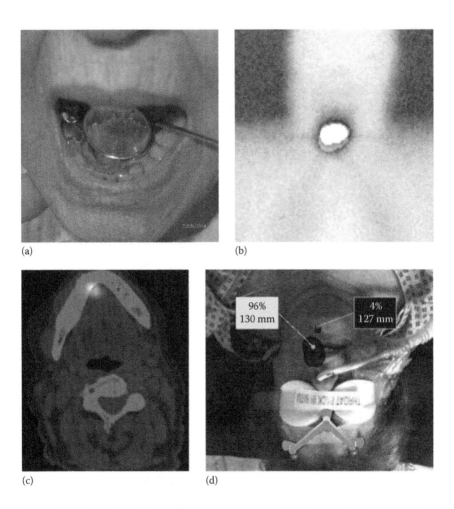

(a)   (b)

(c)   (d)

FIGURE 8.2   Shine-through effect. (a) Clinical photograph of oral squamous cell carcinoma of the lower alveolus. (b) Lymphoscintigraphy with no sentinel nodes seen. (c) **(See colour insert.)** SPECT/CT showing injection site only, no SN were seen. (d) fhSPECT showing injection site and sentinel node. (Images courtesy of M. McGurk and C. Schilling; Adapted from Clinical and Translational Imaging.)

8.4 SENTINEL NODE PROTOCOL IN ORAL CANCER

A detailed guidance for performing SNB in OSCC has been published (Alkureishi et al. 2009). Generally, SNB for OSCC can be undertaken in a 1 or 2-day protocol with no difference in the sensitivity of the technique. In the authors' experience, we find that up to 40 MBq in a 2-day protocol and 20 MBq for 1 day is usually sufficient for localisation, and as little as 10 MBq can give very good resolution of the SN whilst reducing the injection site signal. It is extremely infrequent (<1% of cases) that a SN cannot be identified by lymphoscintigraphy (static and dynamic imaging) although there is increasing use of SPECT/CT for obtaining additional anatomical detail.

During surgery, the SN are identified by means of a handheld gamma probe, and many surgeons will also use an optical tracer in the form of blue dye injected around the tumour (Patent Blue V in Europe or isosulfan blue in the United States). Blue dye can help in the detection of nodes that have already been localised by the gamma signal, but there are disadvantages such as staining at the tumour site, rapid transit of the dye through the nodes and that the colour can be confused with venous structures.

Once the nodes have been excised, they are listed according to gamma count, the presence of blue dye, neck level and size. The radiation in the nodal bed after each node is removed is recorded to ensure no hot nodes are retained, and the background radiation is also reported. An SN should be more than 10 times the background radiation and any node with radioactivity less than 10% of the hottest node is not considered an SN. It is usual to remove 2–3 SN but this can be increased in midline cases with bilateral drainage.

SNs are sent for detailed histopathological analysis by serial step sectioning and immunohistochemical analysis. This highly accurate system will detect micrometastasis and isolated tumour cells, but it can take several days to complete the protocol. Some centres use frozen section analysis as an on-table tool to decide if completion neck dissection is required, with the remainder of the node submitted for serial section as confirmation. Frozen section however is only effective if a sizeable metastatic deposit is present in the section, and with reported accuracy as low as 50%, most centres do not use it routinely (Bilde et al. 2008, Melkane et al. 2012, Vorburger et al. 2012). This means that patients who have a positive SNB diagnosed will have the completion neck dissection as a staged procedure, usually in the 2–3 weeks following the original surgery.

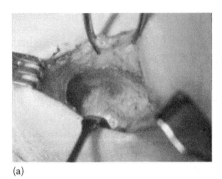

(a)   (b)

FIGURE 8.3   (**See colour insert.**) (a) Photograph of a lymph node stained with blue dye as seen during surgery. (b) Ex vivo recording gamma counts using a handheld gamma probe. (Courtesy of M. McGurk.)

New pathological techniques such as one step nucleic acid amplification (Goda et al. 2012) and quantitative real-time reverse transcriptase polymerase chain reaction (Ferris et al. 2011) offer the prospect of on-table diagnosis of micrometastasis, but this approach is yet to be comprehensively validated in OSCC (Figure 8.3).

8.5 USE OF HYBRID TRACERS IN ORAL CANCER

The development of a multimodal tracer combining the properties of ^{99m}Tc and indocyanine green is described elsewhere in this book. This hybrid tracer has been successfully used for oral cancer and in fact may be most relevant in this area, particularly in relation to tumours of the FOM where there is a high signal to background ratio due to proximity of the injection site (van den Berg et al. 2012). Unlike the traditional optical tracer (blue dye), fluorescence has some penetration through the overlying tissues reducing the amount of tissue dissection required to identify the SN. Moreover the excitation and emission spectra of ICG does not overlap with any tissue autofluorescence resulting in a high specificity for the labelled node. Interestingly, blue dye exhibits autofluorescence but at a lower wavelength than ICG spectra, and there have been reports exploiting this phenomenon for SNB (Tellier et al. 2012). The major advantage of the hybrid tracer over blue dye is that the stable compound is retained within the SN by macrophage phagocytosis, resulting in high level of concordance of fluorescent and gamma signal even in a 2-day protocol; however, there is more work to be done to validate that this is a large-scale study in OSCC.

One of the difficulties using fluorescent image-guided surgery (FIGS) in head and neck region is the interference of ambient theatre light. This is in contrast to minimal access body cavity surgery – for example, laparoscopic prostatectomy – where light contamination is much reduced by nature of the surgical access. At the present time, most head and neck surgery is performed open, and although there is some interest in transaxillary robotic surgery for neck dissection, this is very unlikely to become a standard approach to the neck in the near future.

Ideally, the addition of FIGS to the SN procedure would not interfere with the surgical flow and so integrated theatre lights which can be controlled with a single foot pedal, modified LED surgical headlight (Stoffels et al. 2015), near-infrared region (NIR) cameras that are effective in white light (Schols et al. 2015) as well as integrated dual modality probes may be necessary (van den Berg et al. 2015).

Currently, there is much interest in a new tracer, ^{99m}Tc-labelled tilmanocept known commercially as Lymphoseek, which appears to be selectively accumulated into lymphatic macrophages via CD206 mannose receptor–mediated uptake (Azad et al. 2015). The tracer has shown promising results in SN identification from OSCC, with a low FNR of 2.56% validated against concurrent neck dissection (Agrawal et al. 2015). Qin et al. (2013) explored the multimodal potential of Lymphoseek by radiolabelling with ^{68}Ga (instead of ^{99m}Tc) and adding the fluorophore IRDye 800CW. This new compound allows not only intraoperative gamma and fluorescent signal detection but also for preoperative SN localisation by hybrid PET/CT.

Future developments in tumour surgery are likely to include tumour-specific labelled markers for intraoperative margin analysis. It is possible that these may have a role in SNB although due to the small size of tumour deposits the requirement for signal amplification to allow detection would be the limiting factor (Figure 8.4).

8.6 HYBRID IMAGING IN ORAL CANCER

Until recently, routinely used hybrid imaging for OSCC was limited to SPECT/CT but increasingly PET/CT and PET/MRI as well as new developments such as fhSPECT/Ultrasound imaging have shown promise.

Wagner et al. (2004) first described the advantages of using the combined functional and anatomical information provided by fusion of SPECT (single positron emission computed tomography) and conventional CT (SPECT/CT) over planar lymphoscintigraphy for oral cancer. Subsequently,

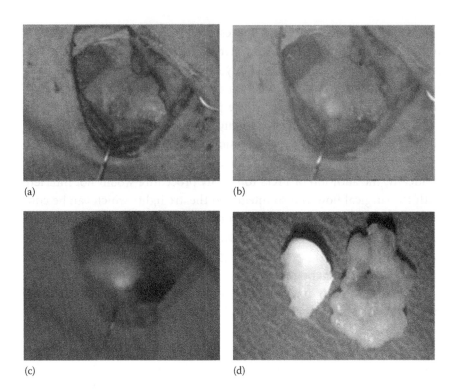

(a)   (b)

(c)   (d)

FIGURE 8.4 **(See colour insert.)** Hybrid ICG and Tc^{99m}-nanocolloid tracer for sentinel node biopsy. (a) Access to sentinel node in level II, sternocleidomastoid muscle (SCM) retracted. (b) Surgical field illuminated with near-infrared region (NIR) and white light showing sentinel node (SN) in front of SCM. (c) Surgical field with NIR light only showing fluorescent SN. (d) Fluorescent signal from sentinel node (left) compared to fibroadipose tissue (right) under NIR light. (Courtesy of M. McGurk and C. Schilling.)

studies showed conflicting results of the utility of SPECT/CT over planar lymphoscintigraphy (Haerle et al. 2009), and although it is not considered essential for accurate SNB (Alkureishi et al. 2009), it remains a popular investigation allowing surgical access planning by anatomical localisation of the SN in relation to critical structures (Stephan and Sandro 2010). The major disadvantage to SPECT/CT is the translatability of the information to the surgical field, particularly once the anatomy has been disturbed during the procedure.

Freehand SPECT (fhSPECT), although not technically a hybrid imaging modality, allows for real-time intraoperative 3D localisation of the SN by algorithmic processing of radiation counts per second in relation to tracking devices placed on the patient and gamma probe. The 3D 'hotspot' map is superimposed onto a video image of the patient allowing tracking of the nodes during surgery with a high degree of spatial accuracy (Figure 8.5). The software is designed to provide fast and repeatable reconstruction of the scanned area, which can be performed throughout the

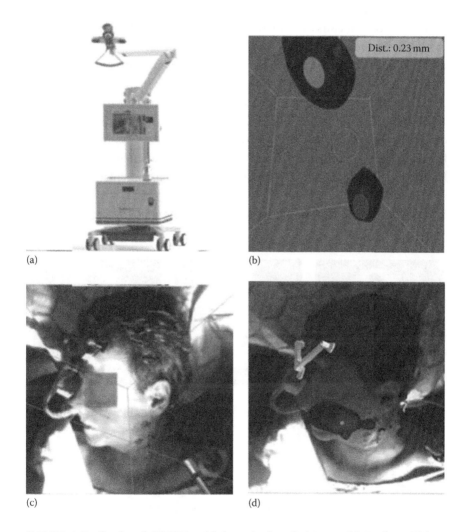

(a)

(b)

Dist.: 0.23 mm

(c)

(d)

FIGURE 8.5   Freehand SPECT-guided sentinel node biopsy. (a) DeclipseSPECT cart system (SurgicEye GmbH, Munich, Germany). (b) fhSPECT in 3D mode showing two sentinel nodes target in centre represents the end of the handheld gamma probe. (c) Patient with fiducial markers on forehead, gamma probe seen in bottom right corner. (d) Same patient after reconstruction of fhSPECT scan showing injection site and three sentinel nodes. (Courtesy of M. McGurk and C. Schilling.)

surgery thus responding to changes in position and anatomy without significant lengthening of the operating time. The utility of the declips-eSPECT system (SurgicEye GmbH, Munich, Germany) has been investigated in OSCC (Bluemel et al. 2014, Mandapathil et al. 2014, Heuveling et al. 2015) with encouraging results. Heuveling et al. (2015) reported that the fhSPECT technique was reliable in 94% of cases and provided a benefit over traditional localisation methods in 24% of cases. In our experience, fhSPECT can be as accurate as SPECT/CT when used in a strict scanning protocol, and this may prove its utility in allowing intraoperative real-time SN injection and localisation thus bypassing the need for imaging studies in previously inaccessible tumours such as the larynx.

Images produced by fusion of PET and sequential or simultaneous MRI or CT scans are useful not only for staging but are increasingly under investigation with respect to biomarkers such as hypoxia, angiogenesis and capillary permeability (Ktrans) in assessing response to treatment and diagnosis of recurrence (Loeffelbein et al. 2012, Pasha et al. 2015). Heuveling et al. (2013) investigated the possibility of a PET/CT lymphoscintigraphy using

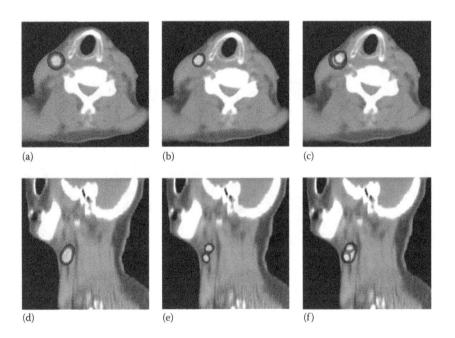

(a)　　　　　　　　(b)　　　　　　　　(c)

(d)　　　　　　　　(e)　　　　　　　　(f)

FIGURE 8.6 **(See colour insert.)** (a–f) Comparison of SPECT/CT (Tc^{99m}-nanocolloid, red) and PET/CT lymphoscintigraphy (^{89}Zr-nanocolloid, blue). (Adapted from Heuveling, D.A. et al., *J. Nucl. Med.*, 54(4), 585, 2013.)

a novel tracer ^{89}Zr-nanocolloid albumin, for five patients with oral cancer who also underwent traditional gamma camera imaging by SPECT/CT using ^{99m}Tc-nanocolloid. They found that PET/CT lymphoscintigraphy not only identified the same SN hotspots as SPECT/CT but also located further SN close to the injection site. Additionally, the authors noted that they were able to identify lymphatic channels draining into the lymph nodes that could not be seen on the SPECT/CT. This advantage however did not translate to the operating room as there is currently no appropriate intraoperative detection device for the ^{89}Zr labelled tracer (Figure 8.6).

A novel fusion imaging device has been proposed by SurgicEye (GmbH, Munich, Germany) combining fhSPECT and ultrasound imaging for identification of SNs and thyroid and parathyroid lesions. This offers the possibility to localise the SN allowing percutaneous sampling by fine-needle aspiration as described by de Bree et al. (2015). Further applications could include direct injection of an optical tracer or placement of permanent radioactive source such as I^{125} seeds which would aid intraoperative localisation of the SN.

8.7 SUMMARY

We are on the cusp of many new developments and collaborations between surgical and imaging specialties that will ultimately improve the accuracy of staging and treatment for oral cancer. The blurring of the continuum between diagnostics and therapeutics alongside technological advances give us the best hope of improving outcomes for our patients. Furthermore, the advances that have been made are applicable to many if not most solid tumour types, and we should aim to share knowledge and experience across specialties for the benefit of patients.

REFERENCES

Agrawal, A., F.J. Civantos, K.T. Brumund et al. 2015. [^{99m}Tc] Tilmanocept accurately detects sentinel lymph nodes and predicts node pathology status in patients with oral squamous cell carcinoma of the head and neck: Results of a phase III multi-institutional trial. *Annals of Surgical Oncology*, 22(11):3708–3715.

Alkureishi, L.W., Z. Burak, J.A. Alvarez et al. 2009. Joint practice guidelines for radionuclide lymphoscintigraphy for sentinel node localization in oral/oropharyngeal squamous cell carcinoma. *European Journal of Nuclear Medicine and Molecular Imaging*, 36(11):1915–1936.

Alkureishi, L.W., G.L. Ross, T. Shoaib et al. 2010. Sentinel node biopsy in head and neck squamous cell cancer: 5-year follow-up of a European multicenter trial. *Annals of Surgical Oncology*, 17(9):2459–2464.

Alvarez, J., A. Bidaguren, M. McGurk et al. 2014. Sentinel node biopsy in relation to survival in floor of the mouth carcinoma. *International Journal of Oral and Maxillofacial Surgery, 43*(3):269–273.

Azad, A.K., M.V. Rajaram, W.L. Metz et al. 2015. γ-Tilmanocept, a new radio-pharmaceutical tracer for cancer sentinel lymph nodes, binds to the mannose receptor (CD206). *The Journal of Immunology, 195*(5):2019–2029.

Bilde, A., C. von Buchwald, M.H. Therkildsen et al. 2008. Need for intensive histopathologic analysis to determine lymph node metastases when using sentinel node biopsy in oral cancer. *The Laryngoscope, 118*(3):408–414.

Bluemel, C., K. Herrmann, A. Kübler et al. 2014. Intraoperative 3-D imaging improves sentinel lymph node biopsy in oral cancer. *European Journal of Nuclear Medicine and Molecular Imaging, 41*(12):2257–2264.

Broglie, M.A., S.K. Haerle, G.F. Huber, S.R. Haile and S.J. Stoeckli. 2013. Occult metastases detected by sentinel node biopsy in patients with early oral and oropharyngeal squamous cell carcinomas: Impact on survival. *Head and Neck, 35*(5):660–666.

Broglie, M.A., S.R. Haile and S.J. Stoeckli. 2011. Long-term experience in sentinel node biopsy for early oral and oropharyngeal squamous cell carcinoma. *Annals of Surgical Oncology, 18*(10):2732–2738.

Burcia, V., V. Costes, J.L. Faillie et al. 2010. Neck restaging with sentinel node biopsy in T1-T2N0 oral and oropharyngeal cancer: Why and how? *Otolaryngology-Head and Neck Surgery, 142*(4):592–597.

Chaturvedi, A.K., W.F. Anderson, J. Lortet-Tieulent et al. 2013. Worldwide trends in incidence rates for oral cavity and oropharyngeal cancers. *Journal of Clinical Oncology, 31*(36):4550–4559.

Civantos, F.J., R.P. Zitsch, D.E. Schuller et al. 2010. Sentinel lymph node biopsy accurately stages the regional lymph nodes for T1-T2 oral squamous cell carcinomas: Results of a prospective multi-institutional trial. *Journal of Clinical Oncology, 28*(8):1395–1400.

de Bree, R., B. Pouw, D.A. Heuveling and J.A. Castelijns. 2015. Fusion of freehand SPECT and ultrasound to perform ultrasound-guided fine-needle aspiration cytology of sentinel nodes in head and neck cancer. *American Journal of Neuroradiology, 36*(11):2153–2158.

Farmer, R.W., L. McCall, F.J. Civantos et al. 2015. Lymphatic drainage patterns in oral squamous cell carcinoma findings of the ACOSOG Z0360 (alliance) study. *Otolaryngology – Head and Neck Surgery, 152*(4):673–677.

Ferris, R.L., L. Xi, R.R. Seethala et al. 2011. Intraoperative qRT-PCR for detection of lymph node metastasis in head and neck cancer. *Clinical Cancer Research, 17*(7):1858–1866.

Flach, G.B., E. Bloemena, W.M.C. Klop et al. 2014. Sentinel lymph node biopsy in clinically N0 T1–T2 staged oral cancer: The Dutch multicenter trial. *Oral Oncology, 50*(10):1020–1024.

Ganly, I., S. Patel and J. Shah. 2012. Early stage squamous cell cancer of the oral tongue – Clinicopathologic features affecting outcome. *Cancer, 118*(1):101–111.

Goda, H., K.I. Nakashiro, R. Oka et al. 2012. One-step nucleic acid amplification for detecting lymph node metastasis of head and neck squamous cell carcinoma. *Oral Oncology*, 48(10):958–963.

Govers, T.M., G. Hannink, M.A. Merkx, R.P. Takes and M.M. Rovers 2013a. Sentinel node biopsy for squamous cell carcinoma of the oral cavity and oropharynx: A diagnostic meta-analysis. *Oral Oncology* 2013;49(8):726–732.

Govers, T.M., W.H. Schreuder, W.M.C. Klop et al. 2016. Quality of life after different procedures for regional control in oral cancer patients: Cross-sectional survey. *Clinical Otolaryngology*, 41(3):228–233.

Govers, T.M., R.P. Takes, B. Karakullukcu et al. 2013b. Management of the N0 neck in early stage oral squamous cell cancer: A modeling study of the cost-effectiveness. *Oral Oncology*, 49(8):771–777.

Haerle, S.K., T.F. Hany, K. Strobel, D. Sidler and S.J. Stoeckli. 2009. Is there an additional value of spect/ct over planar lymphoscintigraphy for sentinel node mapping in oral/oropharyngeal squamous cell carcinoma? *Annals of Surgical Oncology*, 16(11)3118–3124.

Hernando, J., P. Villarreal, F. Alvarez-Marcos, L. Gallego, L. Garcia-Consuegra and L. Junquera. 2014. Comparison of related complications: Sentinel node biopsy versus elective neck dissection. *International Journal of Oral and Maxillofacial Surgery*, 43(11):1307–1312.

Heuveling, D.A., A. van Schie, D.J. Vugts et al. 2013. Pilot study on the feasibility of PET/CT lymphoscintigraphy with 89Zr-nanocolloidal albumin for sentinel node identification in oral cancer patients. *Journal of Nuclear Medicine*, 54(4):585–589.

Heuveling, D.A., S. van Weert, K.H. Karagozoglu and R. de Bree. 2015. Evaluation of the use of freehand SPECT for sentinel node biopsy in early stage oral carcinoma. *Oral Oncology*, 51(3):287–290.

Loeffelbein, D.J., M. Souvatzoglou, V. Wankerl et al. 2012. PET-MRI fusion in head-and-neck oncology: Current status and implications for hybrid PET/MRI. *Journal of Oral and Maxillofacial Surgery*, 70(2):473–483.

Mandapathil, M., A. Teymoortash, J. Heinis et al. 2014. Freehand SPECT for sentinel lymph node detection in patients with head and neck cancer: First experiences. *Acta Oto-Laryngologica*, 134(1):100–104.

Melkane, A.E., G. Mamelle, G. Wycisk et al. 2012. Sentinel node biopsy in early oral squamous cell carcinomas: A 10-year experience. *The Laryngoscope*, 122(8):1782–1788.

Morton, D.L., J.F. Thompson, A.J. Cochran et al. 2014. Final trial report of sentinel-node biopsy versus nodal observation in melanoma. *New England Journal of Medicine*, 370(7):599–609.

Murer, K., G.F. Huber, S.R. Haile and S.J. Stoeckli. 2011. Comparison of morbidity between sentinel node biopsy and elective neck dissection for treatment of the n0 neck in patients with oral squamous cell carcinoma. *Head and Neck*, 33(9):1260–1264.

O'Connor, R., T. Pezier, C. Schilling and M. McGurk. 2013. The relative cost of sentinel lymph node biopsy in early oral cancer. *Journal of Cranio-Maxillofacial Surgery, 41*(8):721–727.

Paleri, V., G. Rees, P. Arullendran, T. Shoaib and S. Krishman. 2005. Sentinel node biopsy in squamous cell cancer of the oral cavity and oral pharynx: A diagnostic meta-analysis. *Head and Neck, 27*(9):739–747.

Pasha, M.A., C. Marcus, C. Fakhry, H. Kang, A.P. Kiess and R.M. Subramaniam. 2015. FDG PET/CT for management and assessing outcomes of squamous cell cancer of the oral cavity. *American Journal of Roentgenology, 205*(2):W150–W161.

Pedersen, N.J., D.H. Jensen, N. Hedbäck et al. 2015. Staging of early lymph node metastases with the sentinel lymph node technique and predictive factors in T1/T2 oral cavity cancer: A retrospective single-center study. *Head and Neck, 38*(S1):E1033–E1040.

Petersen, P.E. 2003. The World Oral Health Report 2003: Continuous improvement of oral health in the 21st century – The approach of the WHO Global Oral Health Programme. *Community Dentistry and Oral Epidemiology, 31*(S1):3–24.

Pezier, T., I.J. Nixon, B. Gurney et al. 2012. Sentinel lymph node biopsy for T1/T2 oral cavity squamous cell carcinoma – A prospective case series. *Annals of Surgical Oncology, 19*(11):3528–3533.

Qin, Z., D.J. Hall, M.A. Liss et al. 2013. Optimization via specific fluorescence brightness of a receptor-targeted probe for optical imaging and positron emission tomography of sentinel lymph nodes. *Journal of Biomedical Optics, 18*(10):101315.

Schiefke, F., M. Akdemir, A. Weber, D. Akdemir, S. Singer and B. Frerich. 2009. Function, postoperative morbidity, and quality of life after cervical sentinel node biopsy and after selective neck dissection. *Head and Neck, 31*(4):503–512.

Schilling, C., S.J. Stoeckli, S.K. Haerle, M.A. Broglie, G.F. Huber, J.A. Sorensen, V. Bakholdt et al. 2015. Sentinel European Node Trial (SENT): 3-year results of sentinel node biopsy in oral cancer. *European Journal of Cancer, 51*(18):2777–2784.

Schols, R.M., N.J. Connell and L.P. Stassen. 2015. Merged near-infrared fluorescence and white light imaging in minimally invasive surgery. *World Journal of Surgery, 39*(8):2106.

Shoaib, T., D.S. Soutar, J.E. Prosser et al. 1999. A suggested method for sentinel node biopsy in squamous cell carcinoma of the head and neck. *Head and Neck, 21*(8):728–733.

Stephan, K.H. and J.S. Sandro. 2010. SPECT/CT for lymphatic mapping of sentinel nodes in early squamous cell carcinoma of the oral cavity and oropharynx. *International Journal of Molecular Imaging, 2011*, 1–6.

Stoeckli, S.J., S.K. Haerle, K. Strobel, S.R. Haile, T.F. Hany and B. Schuknecht. 2012. Initial staging of the neck in head and neck squamous cell carcinoma: A comparison of CT, PET/CT, and ultrasound-guided fine-needle aspiration cytology. *Head and Neck, 34*(4):469–476.

Stoffels, I., J. Leyh, T. Pöppel, D. Schadendorf and J. Klode. 2015. Evaluation of a radioactive and fluorescent hybrid tracer for sentinel lymph node biopsy in head and neck malignancies: Prospective randomized clinical trial to compare ICG-^{99m}Tc-nanocolloid hybrid tracer versus ^{99m}Tc-nanocolloid. *European Journal of Nuclear Medicine and Molecular Imaging, 42*(11):1631–1638.

Tellier, F., J. Steibel, R. Chabrier et al. 2012. Sentinel lymph nodes fluorescence detection and imaging using Patent Blue V bound to human serum albumin. *Biomedical Optics Express, 3*(9):2306–2316.

van den Berg, N.S., O.R. Brouwer, W.M. Klop et al. 2012. Concomitant radio- and fluorescence-guided sentinel lymph node biopsy in squamous cell carcinoma of the oral cavity using ICG-^{99m}Tc-nanocolloid. *European Journal of Nuclear Medicine and Molecular Imaging* 2012;39(7):1128–1136.

van den Berg, N.S., H. Simon, G.H. Kleinjan et al. 2015. First-in-human evaluation of a hybrid modality that allows combined radio-and (near-infrared) fluorescence tracing during surgery. *European Journal of Nuclear Medicine and Molecular Imaging, 42*(11):1639–1647.

Van Leeuwen, A.C., T. Buckle, G. Bendle et al. 2011. Tracer-cocktail injections for combined pre-and intraoperative multimodal imaging of lymph nodes in a spontaneous mouse prostate tumor model. *Journal of Biomedical Optics, 16*(1):016004.

Vorburger, M.S., M.A. Broglie, A. Soltermann et al. 2012. Validity of frozen section in sentinel lymph node biopsy for the staging in oral and oropharyngeal squamous cell carcinoma. *Journal of Surgical Oncology, 106*(7):816–819.

Wagner, A., K. Schicho, C. Glaser et al. 2004. SPECT-CT for topographic mapping of sentinel lymph nodes prior to gamma probe-guided biopsy in head and neck squamous cell carcinoma. *Journal of Cranio-Maxillofacial Surgery, 32*(6):343–349.

Warnakulasuriya, S. 2009. Global epidemiology of oral and oropharyngeal cancer. *Oral Oncology, 45*(4):309–316.

CHAPTER 9

Characterisation and Quality Assurance Protocols for SFOV Gamma Cameras

S.L. Bugby, John E. Lees and Alan C. Perkins

CONTENTS

9.1 INTRODUCTION

Small field of view (SFOV) gamma cameras are now used in a number of diagnostic and interventional procedures (Perkins and Hardy 1996, Duch 2011). As a medical tool, it is vitally important that the performance of these systems can be tested in a quantitative and repeatable way.

Standardised procedures for assessing the performance characteristics of medical gamma cameras have generally been based on the original standards published by the U.S. National Electrical Manufacturing Association (NEMA) (Chapman et al. 2007). The European Directive 97/43/EURATOM mandates a quality assurance programme for all medical devices used in diagnostic radiology, nuclear medicine and radiotherapy (Teunen 1998). In the United Kingdom and Europe, a comprehensive description of procedures to be carried out in clinical departments has been developed by the Institute of Physics and Engineering in Medicine (IPEM) (IPEM 2003). However, these tests are designed for use with standard large field of view (LFOV) gamma cameras and are not necessarily applicable to SFOV systems.

We have proposed a generic scheme suitable for evaluating the performance characteristics of SFOV gamma cameras (Bhatia et al. 2015), based on modifications to the standard procedures described by NEMA (NU1 2007). The key differences in methodology between tests for LFOV and SFOV gamma cameras are described. It is envisaged that this scheme will provide more appropriate methods for equipment characterisation, ensuring quality and consistency for all SFOV cameras. In this chapter, we summarise the salient issues – a more complete description of the SFOV assessment scheme can be found in Bhatia et al. (2015).

9.2 UNIQUE CHALLENGES IN CHARACTERISING SFOV GAMMA CAMERAS

The term 'SFOV gamma camera' is ambiguous and has been applied to cameras with field of view (FOV) as large as 400 mm × 400 mm or as small as 20 mm × 20 mm. For cameras at the higher end of this range, the IPEM characterisation procedures are appropriate and can be used without adaptation. For instruments operating towards the lower end, particularly those with a FOV smaller than 50 mm × 50 mm, these standards are less applicable and in some cases may not be possible to perform.

One common tendency in SFOV gamma camera design is a move away from the bulky photon multiplier tube (PMT) detectors used in standard gamma cameras. SFOV cameras may use more compact position sensitive PMTs, photodiodes, CMOS or CCD detectors coupled to a scintillating material, or may directly detect gamma photons with high-Z semiconductor materials such as CdTe or CZT. These newer detectors can improve on the spatial resolution achievable with a LFOV camera by a factor of up to 10.

Due to the need for portability, and sometimes due to the design constraints of a chosen detector, the detectors in SFOV cameras tend to be far smaller than in their LFOV counterparts. If a parallel hole collimator – the typical choice for LFOV cameras – is used, then the FOV of the camera is limited to the size of the detector.

Conversely, images taken with a pinhole collimator can be magnified or de-magnified depending on the imaging configuration used. The magnification of a pinhole collimator is

$$M = \frac{t}{h} \tag{9.1}$$

where

  t is the distance between the collimator (or, more precisely, the centre of the pinhole within the collimator) and the detector (usually a fixed parameter)

  h is the distance between the collimator and the source being imaged

This magnification gives pinhole collimated cameras a variable FOV which can be many times larger than the detector size and may be adjusted by varying the distance between the camera and the source being imaged. SFOV cameras often make use of pinhole collimators to counteract their smaller detector size.

Each of these differences in camera design has implications for appropriate characterisation procedures as discussed in the following text. In addition, the variation in design of SFOV cameras is far greater than that of LFOV cameras currently in use. For this reason, these SFOV protocols have been designed with flexibility in mind, so that they may be adapted for a range of systems without sacrificing the ease of comparison between measurements from different systems.

9.2.1 Detector Size

Measurements of intrinsic properties, such as spatial resolution or linearity, are more difficult with small detectors as the test materials must be constructed with greater precision than would be needed for a larger detector.

For example, for LFOV cameras, intrinsic spatial resolution is measured with a capillary line source. For accurate measurements, it is advised that the source has an internal diameter of no larger than 80% of the expected resolution. LFOV cameras have typical intrinsic resolutions of the order 3 mm, and line sources with an internal diameter of 0.5 mm are typically used (IPEM 2003).

The reported intrinsic spatial resolution of SFOV cameras can be as low as 100 μm (de Vree et al. 2005). Following the standards for LFOV cameras would therefore require a line source with a diameter less than 80 μm. At these scales, the difficulty in manufacture and filling of phantoms without specialist equipment becomes a limiting factor (Vastenhouw and Beekman 2007, Lees et al. 2010), and so an alternative method has been proposed for SFOV measurements.

Traditional phantoms designed for LFOV spatial linearity measurements, such as the parallel line equal spacing (PLES) phantom, are vastly oversized for SFOV systems. A typical PLES phantom may consist of a series of parallel 1 mm width line features separated by 20 mm, and around 50 line features would be expected in a single LFOV image. When used with a SFOV camera, this separation may mean only one or two features are visible on the resultant image and the scale of the lines compared to the detector means that their width can no longer be considered negligible.

To achieve a comparable image to that obtained from a LFOV system, a SFOV PLES phantom would need to be scaled down by a factor of up to 100 which would require precise manufacturing beyond the scope of most nuclear medicine departments. Unlike phantoms for LFOV systems, which are readily available from a number of manufacturers, standardised phantoms for SFOV cameras are yet to become commercially available.

A number of LFOV characterisation protocols specify total photon counts in terms of detected photons in order to ensure a statistically representative image. For smaller detectors, the total required number of counts may be reduced while maintaining the same statistical uncertainties as would be expected for LFOV cameras. Due to the wide range of detector designs used in SFOV cameras, stop conditions should be flexible but the application of these must remain straightforward.

Despite these difficulties, smaller detector sizes do reduce the complexity of some required tests. When a uniform source is required for a test on a

LFOV detector, a flood source is typically used – this may be a solid source with associated difficulties for storage and handling or a liquid source which tends to be heavy and difficult to fill uniformly. An alternative to a flood source, for intrinsic tests, is to place a point source at a sufficient distance from the detector to produce a uniform flux. To reduce flux non-uniformity to less than 1% requires $h > 5D$, where h is the source–detector distance and D is the largest dimension of the detector. For a LFOV camera, this would require a point source to be placed several metres distant from the detector and, as the intensity of a source drops off with h^2, the activity within the source would have to be significant. For SFOV cameras, the uniformity condition may be satisfied at distances as small as 50 mm, making appropriate shielding far more manageable and so making the point source method more convenient than the use of a flood source.

9.2.2 Pinhole Collimation

For LFOV cameras, performance characteristics such as system resolution are stated 'at the camera face', that is, with a source placed directly on top of the collimator, the closest possible approach to the camera. For a pinhole collimator, a source at the collimator surface is not a useful measurement point as any source would appear as a flood source in the resulting image.

System spatial resolution varies with source-collimator distance. As the vast majority of LFOV cameras are parallel hole collimated, operators can be expected to be experienced in translating the reported parameters – typically spatial resolution at the camera face and at a distance of 10 cm – into an understanding of camera performance at any distance, whereas this might not be the case for pinhole collimators. The sensitivity of a parallel hole collimated system is invariant with source–camera distance. This is not the case for pinhole collimators and so a single measurement of extrinsic sensitivity is no longer sufficient. Additionally, due to the variable FOV and smaller common detector size for SFOV pinhole collimated cameras, the distance between the source and camera for different imaging tests is likely to vary far more than for LFOV cameras. For these reasons, it is recommended to provide more information when characterising these cameras, ideally enough that an operator would be able to calculate the expected spatial resolution and sensitivity for any source–camera distance using only this information.

Pinhole collimators also introduce spatial non-uniformity to images and so an extrinsic uniformity measurement should be taken in place of the standard intrinsic uniformity test.

9.2.3 Improved Spatial Resolution

Spatial resolution is measured using a point or line source – these are sources that are small enough that their width can be considered negligible when compared to the spatial resolution of the camera. For LFOV systems, a line or point source with a diameter of around 0.5 mm is used – an appropriate size when spatial resolution is expected to be of the order of 5 mm but far too large to be considered negligible when assessing systems that may have sub-mm resolutions. The smaller line or point sources required for SFOV high-resolution measurements are difficult to manufacture and fill (Lees et al. 2010).

It may be possible to use a point or line source of a known size and then deconvolve the expected profile from the resultant image to determine resolution; however, this requires specific knowledge of the expected profile of the source which is not necessarily a trivial task, particularly for pinhole collimators, and so is not an appropriate method in most cases.

To solve this problem, the use of a point source as small as can reasonably be filled is recommended. Resolution measurements are approximately constant for any source size smaller than the geometric resolution of the pinhole collimator (Bugby et al. 2012), given by

$$R = d\left(\frac{1}{M} + 1\right) \tag{9.2}$$

where

M is the magnification associated with the camera and source positions
d is the diameter of the pinhole

From Equation 9.1, the appropriate minimum source distance can then be chosen for a given source size R. The relationship between source distance and resolution is known to be linear when intrinsic resolution is small compared to expected extrinsic resolution. Therefore, a series of resolutions at larger source distances may be extrapolated to those at smaller distances – which may be more clinically useful – without the need to construct and fill very small sources.

9.3 CHARACTERISATION PROTOCOL FOR SFOV CAMERAS

A characterisation protocol for SFOV cameras is outlined in the following text. These protocols have not been defined in their entirety and more detail may be found in Bhatia et al. (2015) with a full example characterisation available in Bugby et al. (2014).

For all tests, the manufacturer's standard operating procedures should be used. This includes any energy windowing (the specifics of which should be stated in the characterisation) and uniformity correction. Many SFOV cameras use a smoothing algorithm to improve the interpretability of low count images and this should be applied prior to analysis in characterisation images. All measurements should be performed separately for each radioisotope which will be used with the camera.

The appropriate activity of each source used depends on the count rate capability and sensitivity of the camera. The correct positioning and size of sources will also depend on camera design.

9.3.1 Intrinsic Spatial Resolution

Intrinsic spatial resolution is defined as the full width at half maximum (FWHM) of a line spread function (LSF) or point spread function (PSF) without an imaging collimator installed. A full width at tenth maximum (FWTM) measurement should also be stated as the PSF or LSF may deviate from a Gaussian profile (IPEM 2003).

In these protocols, intrinsic spatial resolution is calculated using the edge response function (ERF) method which eliminates the need for very small sources.

A mask with a machined edge, manufactured from a material with low transmission for the photon energies being used, is required. When irradiated with a uniform radioactive source, such that incident gamma photons can be assumed to be perpendicular to the mask plane, this will produce an image of the edge mask, similar to the simulated image in Figure 9.1. There will be a high count region, where the mask did not cover the detector, a low count region, where the mask covered the detector, and a roll off in counts between the two which is related to the spatial resolution of the detector.

To analyse this image an ERF must be created. For an edge mask image, the ERF created by plotting the position on the image plane (perpendicular to the mask edge) against normalised counts (see Figure 9.1) approximates a step function. The ERF is then differentiated to give a LSF, also shown in Figure 9.1. Intrinsic spatial resolution should be reported as the mean FWHM and mean FWTM of the LSF.

9.3.2 System Spatial Resolution

System, or extrinsic, spatial resolution is the FWHM of a LSF or PSF measured when the imaging collimator is in place.

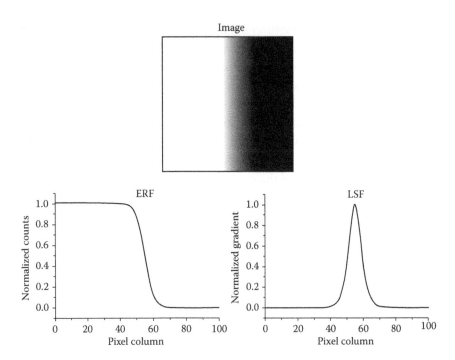

FIGURE 9.1 A simulated edge mask image with the resultant edge response function and line spread function graphs shown.

For pinhole collimators, the maximum source width should be limited by the geometric resolution of the pinhole at the closest distance to be tested. For parallel hole collimators, a source with a maximum diameter of 0.5 mm should be used, and imaging should be carried out at distances such that the expected spatial resolution is at least 50% higher than the size of the source.

The test source should be imaged over at least five distances, ideally covering the range of distances expected in operation. This test should be conducted both with and without scattering material between the source and the camera. For each image, the FWHM and FWTM of the source should be calculated and these values plotted on a graph. For pinhole collimators, the measured resolutions (on the detector) must be corrected for the magnification of the pinhole.

For both parallel and pinhole collimators, the relationship between spatial resolution and source distance is expected to be linear and so a line of the form

$$R = Ah + b \tag{9.3}$$

can be fitted to the plotted graph. The fitted coefficients, A and b, should be reported and will allow an operator to calculate resolution R for any source distance h both with and without scattering material. If a single spatial resolution value is required, it should be the resolution at the camera face for parallel hole collimators and at the non-magnifying point for pinhole collimators. Within the characterisation documentation, the ranges over which measurements were taken should also be stated along with the scattering material used.

9.3.3 Intrinsic Spatial Linearity

Spatial linearity is a measure of how accurately event positions are mapped to the resulting image. This is quantified by investigating the difference between the image of a line source and a true straight line.

To measure linearity, a line source must be imaged repeatedly across the detector face. A minimum of five images should be taken across both axes of the camera (10 images in total), although more may be appropriate depending on the detector size. The LFOV method of using a PLES phantom or similar may be used if an appropriately sized transmission mask can be created and intrinsic resolution is sufficiently small. This mask will be imaged in both orientations and each line feature analysed separately. If this is not possible, a single-slit transmission mask may be used, the width of which does not need to qualify as a line source, and imaged repeatedly at different positions over the detector.

For each row or column of the line image (depending on the orientation of the line), the centre-point of the line should be calculated by fitting a rectangular function, if the line source is broad compared to the intrinsic spatial resolution, or a Gaussian function, if the line source is small compared to intrinsic spatial resolution.

The centre point of each line should be plotted against its position on the detector and a linear fit performed to the data. In line with LFOV protocols, absolute linearity (maximum deviation of the data from this fitted line) should be stated and the differential linearity (mean and standard deviation from line calculated across the entire FOV). Another useful parameter is the R^2 fit statistic, which may also be reported as the mean R^2 for all lines.

9.3.4 Spatial Uniformity

Uniformity is a measure of variations in camera detection performance across the detector face. This may be tested both with and without an imaging collimator, with extrinsic uniformity the more clinically valuable quantity and intrinsic uniformity the better parameter for direct comparisons between detectors. Intrinsic uniformity is also a useful measure for a system with a range of possible collimators.

Uniformity measurements are carried out with either a point source (at sufficient distance for incident photons to be considered uniform) or a flood source. In each case the variation in flux from the source should be less than 1%. The activity used and appropriate acquisition time will vary for different cameras. They should be selected so that at least 10,000 counts must be acquired in each pixel over the course of the image to reduce statistical variation to 1%.

There are a number of quantitative measures of spatial uniformity. For SFOV cameras, the suggested reporting parameters are the coefficient of variation (CoV):

$$\text{CoV} = \frac{\sigma}{\mu} \times 100\% \tag{9.4}$$

where μ and σ are the mean and standard deviation in counts across the image. CoV is a measure of integral uniformity as it covers the entire image. A different measure of uniformity can be calculated using

$$U = \left[\frac{C_{max} - C_{min}}{C_{max} + C_{min}} \right] \times 100\% \tag{9.5}$$

where C_{max} and C_{min} are the maximum and minimum number of detected counts per pixel in a block of pixels. This is a more common technique for LFOV characterisation; however, it is sensitive to outlying pixels, with either high or low values, and so CoV is the recommended technique for integral measurements (IPEM 2003).

Differential uniformity is another important measure; in this case, the aim is to quantify local variations in camera performance rather than over the entire imaging area as for integral uniformity. For this test, the detector should be subdivided into at least 20 blocks, although the appropriate number and size of each section will depend strongly on detector design. The uniformity of each section is then calculated using Equation 9.5. The mean and standard deviation of uniformity across the image can then be reported as a measure of differential uniformity.

9.3.5 Sensitivity

Sensitivity is the proportion of photons incident on a camera which are detected. This should be stated both without a collimator (intrinsic) and with a collimator in place (system or extrinsic).

For both intrinsic and system sensitivities a point source of known activity is required. This is imaged over a known period of time, and the total number of counts in the image is recorded.

For intrinsic measurements, sensitivity should be reported as the ratio of detected to incident counts to allow for easy comparison between detectors of different sizes. Incident counts are calculated using solid angle formulae – assuming an isotropic source centred at a distance h from a rectangular detector of dimensions $\alpha \times \beta$, the proportion of generated counts incident on the detector will be

$$\Omega = \frac{1}{\pi} \tan^{-1} \left[\frac{\alpha\beta}{2h\sqrt{4h^2 + \alpha^2 + \beta^2}} \right] \tag{9.6}$$

System sensitivity should be reported in cps/MBq. For parallel hole collimators, a measurement should be made at the collimator surface and at 100 mm distance, the latter both with and without scattering material between the source and camera. For pinhole collimators, measurements should be made at least five distances (both with and without scattering material) and a plot between distance and sensitivity produced. If a single value is required for comparison purposes, the sensitivity at the non-magnifying point should be used.

9.3.6 Count Rate Capability

Count rate capability is a measure of the range over which the detector response to incoming photons is linear. Both intrinsic and extrinsic measurements are possible.

A point source of activity high enough to saturate the detector should be placed at a known distance from the detector; multiple images should be taken as the source decays with acquisition time recorded for each image. Alternatively, multiple sources with differing activities can be used. At least 10 data points are required, covering the full range of detector capability – from a completely saturated image to one where very few counts are detected.

Using Equation 9.6 and the activity of the source during each period of imaging, a plot can be created of incident counts against recorded counts. The count rate capability is expressed as the incident count rate at which the observed count rate is 20% less than would be expected from extrapolating the linear portion of the graph. For a better comparison between different detector types, it may also be useful to provide a value of count rate capability per unit area, as this separates the effect of detector size from the effect of detector design.

9.3.7 Energy Resolution

Energy resolution is an important parameter as it indicates the ability of a system to discriminate true events from scattered events and so improve the spatial resolution and signal to noise ratio of images, or discriminate between photon energies for dual radionuclide tests.

Energy spectra are accumulated using a point source of activity within the count rate capability of the camera. A Gaussian curve is fitted to the photo-peak of the principal emission energy of the radionuclide being imaged. The energy resolution of the camera is then reported as the FWHM of this curve.

9.4 SFOV SPECIFIC TESTS

The tests described in Section 9.3 directly follow from those established for LFOV gamma cameras. The only changes that have been made are those intended to make these tests easier to perform with SFOV systems, without fundamentally changing the method or the purpose of each test. The benefit of this is that these protocols allow direct comparisons to be made between SFOV and LFOV cameras which may be particularly useful in the early days of SFOV camera commercialisation. There is a risk, however, that these tests cannot take into account the difference in uses between SFOV and LFOV camera operation and may not be as relevant to SFOV applications as they are to LFOV applications.

SFOV cameras are often designed for specific purposes such as sentinel lymph node biopsy, parathyroid gland surgery and radioimmunoguided surgery. In these procedures, radiation sources are likely to be superficial and so scatter discrimination becomes a less important feature. Sensitivity remains important, particularly as surgical (and therefore imaging) times must be minimised as much as possible. The importance of spatial resolution will depend strongly on the procedure being carried out – the likelihood of needing to discriminate between spatially close sources is far higher in a head and neck sentinel lymph node biopsy, for example, than it is in a melanoma sentinel lymph node biopsy. More portable SFOV cameras are likely to find their way into the hands of someone who is not a nuclear medicine specialist, as would be the case in surgical use. It is therefore important that reported performance characteristics are intuitive and relate directly to the real-world experiences of the end user.

One possible route for specialised SFOV camera tests would be the use of readily available or easily produced phantoms which aim to simulate medical procedures. A custom-built phantom designed to simulate sentinel lymph node imaging has been developed which is cheap and relatively simple to produce (Ng et al. 2015). Alternatively, sentinel lymph node and vessel phantoms may be assembled using readily available materials such as butterfly tubing and Eppendorf tubes (Alqahtani et al. 2015).

These techniques provide a useful tool for demonstrating the capability of a camera for a specific procedure, but experimental set ups would have to be fully standardised before they could be used for comparison between different imaging systems.

Another possibility is the use of qualitative performance characteristics. One possible candidate for this technique is detectability – the ability of an operator to distinguish a feature from its surroundings. A number of parameters exist for quantifying detectability, one of the most popular being contrast-to-noise ratio, but none of these can definitively say whether an operator would be able to detect a particular feature. Quantitative measures also cannot take into account operator experience and knowledge which can play a large part in determining the existence of features. Detectability will vary between operators and between display methods even when identical data are presented, and so a qualitative investigation covering many of these variables may provide the most clinically useful measure of detectability.

As SFOV cameras continue to be developed, the need for further tests will arise. Some SFOV cameras, such as hybrid gamma camera (HGC), optical-gamma hybrid camera, discussed in Chapter 10 have capabilities that are not present in LFOV cameras and so will never be fully described by traditional characterisation protocols. The alignment of the optical and gamma images in the HGC, for example, must be tested. Whether the results of this testing would prove useful to the end user or are used simply as a form of quality control is yet to be decided, and will remain so until the camera has been in clinical use for some time.

As the use of SFOV cameras continues to expand into different procedures, characterisation tests are likely to also change and expand.

When designing a new test, it is important to ensure that it is reproducible on as wide a range of systems as possible as this is the first step towards standardising in the future.

9.5 QUALITY ASSURANCE TESTING

The performance of a gamma camera may change over time and with use, and it is important that users are aware of the most accurate characterisation data possible for their system. The need for up to date characterisation however must be weighed against the time it would take to regularly perform a complete characterisation. The characterisation protocols described in Section 9.3 may be carried out when a new camera is delivered to a centre or to evaluate a new camera design but would be impractical as a daily test of camera performance. Instead, it is advisable that users implement a quality assurance (QA) scheme, where parameters are regularly checked but to a lesser accuracy than is required for a full characterisation protocol.

Standard quality assurance measures for LFOV cameras vary from centre to centre. IPEM suggests using performance characterisation tests for QA and has published a series of recommended tests and appropriate time intervals for this (IPEM 2003).

QA testing for LFOV cameras often includes intrinsic measurements. The vast majority of LFOV cameras are purchased with a variety of collimators and the removal and replacement of these is a regular task in nuclear medicine departments. Some SFOV cameras, however, are designed for use with only a single collimator and, when this is the case, it may be that the collimator has not been designed for easy removal. For this reason, the QA methods suggested here are all system tests conducted with the collimator in place.

The smaller size of SFOV gamma cameras has allowed some manufacturers (e.g. CrystalCam, Sentinella) to develop and provide jigs which allow for easy and repeatable quality assurance tests to be carried out. Where possible, these jigs should be used along with comparable source activities for each test to provide the best consistency in measurements.

9.5.1 Suggested SFOV QA Scheme

The following proposed scheme aims to provide a general framework, although the need for particular tests will vary depending on detector type. For example, a regular quick test for hot pixels may be appropriate for a CCD detector but will not be applicable to a PSPMT array. Therefore, these protocols should be expanded or adapted based on the camera design chosen. Appropriate time periods between tests may also vary; those suggested in the following text should be considered lower estimates of the required regularity.

In some cases, image degradation may appear between one test and the next, for example, if a collimator has been dropped or damaged. It is also possible that deviation from characterisation parameters may evolve as a drift over several weeks or months. It is therefore preferable to store and track quantitative results from each test. These can be used to determine whether there is a trend away from the reported performance of the camera.

Daily tests

- Visual examination for damage and defects

- Background level – An image with no source present

Weekly tests

- System uniformity (the image required does not need to be as statistically accurate as that for characterisation; 100 photons/pixel should be sufficient)

- Photopeak position

Monthly tests

- Sensitivity

Yearly tests

- Spatial resolution

- Recalibration of uniformity map and photopeak

9.6 SUMMARY AND CONCLUSIONS

The development and a production of SFOV gamma cameras is a growing field. These cameras use many types of detectors, have a wide range in FOV size and collimator design and may be optimised with specific procedures in mind. The huge variety in SFOV camera design makes the production of performance assessment protocols that are applicable to all systems a difficult task although it is necessary to do so in order for these systems to be fairly compared to one another.

The protocols outlined in this chapter aim to balance practical necessities for use with the need for comparability between systems. Although criteria for measurements have been provided, variation in camera design is allowed for by introducing flexibility to the experimental set-ups used to obtain these results. It is hoped that these protocols provide a standardised base for the performance characterisation of SFOV cameras in the future.

The use of SFOV cameras in a clinical setting is still relatively new. As practitioners in the field become more experienced with these devices, a range of new tests may be developed which more directly relate to the performance of the cameras in clinical use. QA protocols will also be further developed and customised for the cameras being tested.

REFERENCES

Alqahtani, M.S., J.E. Lees, S.L. Bugby, L.K. Jambi and A.C. Perkins. 2015. Lymphoscintigraphic imaging study for quantitative evaluation of a small field of view (SFOV) gamma camera. *Journal of Instrumentation*, 10(07): P07011.

Bhatia, B.S., S.L. Bugby, J.E. Lees and A.C. Perkins. 2015. A scheme for assessing the performance characteristics of small field-of-view gamma cameras. *Physica Medica*, 31(1):98–103.

Bugby, S.L., J.E. Lees, B.S. Bhatia and A.C. Perkins. 2014. Characterisation of a high resolution small field of view portable gamma camera. *Physica Medica*, 30(3):331–339.

Bugby, S.L., J.E. Lees and A.C. Perkins. 2012. Modelling image profiles produced with a small field of view gamma camera with a single pinhole collimator. *Journal of Instrumentation*, 7(11):P11025.

Chapman, J., J. Hugg, J. Vesel et al. 2007. *Performance Measurement of Scintillation Cameras*. Rosslyn, VA: National Electrical Manufacturers Association.

de Vree, G.A., A.H. Westra, I. Moody, F. Van der Have, K.M. Ligtvoet and F.J. Beekman. 2005. Photon-counting gamma camera based on an electron-multiplying CCD. *IEEE Transactions on Nuclear Science*, 52(3):580–588.

Duch, J. 2011. Portable gamma cameras: The real value of an additional view in the operating theatre. *European Journal of Nuclear Medicine and Molecular Imaging*, 38(4):633–635.

IPEM. 2003. A. Bolster (Ed.) Quality control of gamma camera systems. Report No. 86. York, UK: Institute of Physics and Engineering in Medicine.

Lees, J.E., D.J. Bassford, P.E. Blackshaw and A.C. Perkins. 2010. Design and use of mini-phantoms for high resolution planar gamma cameras. *Applied Radiation and Isotopes*, 68(12):2448–2451.

Ng, A.H., D. Clay, P.E. Blackshaw et al. 2015. Assessment of the performance of small field of view gamma cameras for sentinel node imaging. *Nuclear Medicine Communications*, 36(11):1134–1142.

NU1. 2007. National Electronics Manufacturer's Association, Rosslyn, VA.

Perkins, A.C. and J.G. Hardy. 1996. Intra-operative nuclear medicine in surgical practice. *Nuclear Medicine Communications*, 17(12):1006–1015.

Teunen, D. 1998. The European Directive on health protection of individuals against the dangers of ionising radiation in relation to medical exposures (97/43/EURATOM). *Journal of Radiological Protection*, 18(2):133.

Vastenhouw, B. and F. Beekman. 2007. Submillimeter total-body murine imaging with U-SPECT-I. *Journal of Nuclear Medicine*, 48(3):487–493.

CHAPTER 10

Hybrid Gamma-Optical Imaging

John E. Lees, S.L. Bugby and Alan C. Perkins

CONTENTS

10.1 INTRODUCTION

A number of high-resolution small field of view (SFOV) cameras have been developed for clinical use (Tsuchimochi and Hayama 2013, Olcott et al. 2014) enabling imaging procedures to be undertaken at the bedside, in intensive care units and clinics, and in the operating theatre (Perkins and Hardy 1996, Duch 2011).

Following the phenomenal success of positron emission tomography–computed tomography (PET-CT), there is a growing interest among researchers, clinicians and manufacturers in combining imaging systems from different modalities into single multipurpose hybrid devices. These hybrid cameras combine functional nuclear imaging with anatomical imaging, providing improved localisation of sources; however, the surgical use of these devices is limited by their large size and immobility.

This chapter describes the Hybrid Gamma Camera (HGC), a SFOV portable system for hybrid gamma and optical imaging, designed for small-organ and intraoperative applications. This may allow for the more intuitive interpretation of SFOV gamma images and an improvement in accuracy of the localisation of areas of tracer uptake. Results from phantom studies and initial clinical evaluations using the prototype HGC system are presented.

10.2 HYBRID IMAGING IN NUCLEAR MEDICINE

Although gamma imaging provides functional information unavailable with alternative imaging modalities, there may be difficulties in relating the distribution of uptake to anatomical sites; hence, the interpretation of images requires the skills of a highly trained nuclear medicine specialist. One solution to assist in interpretation is multimodal or hybrid imaging – the simultaneous acquisition of multiple imaging types – which has now become common practice in nuclear medicine.

PET images are now invariably taken in combination with an x-ray CT scan (Morris and Perkins 2012). The CT provides anatomical information and allows for attenuation correction, while PET is used to analyse functional or metabolic processes. Single photon emission computed tomography (SPECT), the 3D implementation of the gamma imaging camera, is now also routinely combined with CT (Zaidi and Prasad 2009, Visioni and Kim 2011). The success of these techniques has led to the further development of hybrid systems using PET and magnetic resonance imaging (MRI) with a new generation of PET-MRI devices now in clinical use (Cherry 2009, Beyer et al. 2011).

Over recent years, the major imaging equipment manufacturers have been advancing the capabilities of the large hybrid PET-CT, SPECT-CT and PET-MRI scanners. These devices can speed up diagnostic procedures by collecting all the necessary information simultaneously or sequentially rather than sending the patient for separate diagnostic investigations. By their very nature, these systems are large and expensive and normally housed in specially constructed environments which are managed and operated by highly trained staff.

In contrast to this trend towards large dedicated facilities, a number of researchers have been developing compact gamma cameras offering greater flexibility for staff and patients. Until recently, all of these smaller systems have focussed on a single modality.

10.3 HYBRID GAMMA CAMERA

The HGC prototype has been developed at the Space Research Centre, University of Leicester, in collaboration with Nottingham University Hospitals Trust and the University of Nottingham. The HGC (Figure 10.1) combines an optical and a gamma detector in a co-aligned configuration for high spatial resolution hybrid imaging. The HGC may be used as a handheld device or mounted on an arm to allow hands free imaging during surgery.

10.3.1 Camera Design

The HGC gamma detector is an electron multiplying charge coupled device (e2v CCD97 BI [e2v Technologies Ltd, Essex, UK]) coupled to a 1500 μm thick columnar CsI:Tl scintillator (Hamamatsu Photonics UK Limited, Hertfordshire, UK) with an imaging area of approximately 8 mm × 8 mm. The calculated absorption of the scintillator is 100%–40% over an energy range of 30–140 keV.

A pinhole collimator allows the imaging field of view (FOV) to vary depending on imaging distance with a minimum FOV of 20 mm × 20 mm

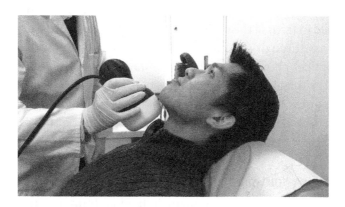

FIGURE 10.1 Photograph of hybrid gamma camera being used in a clinical setting.

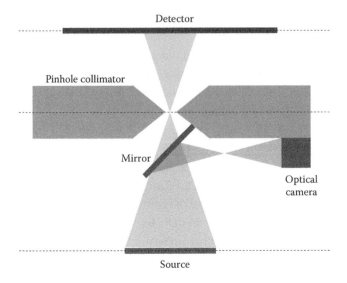

FIGURE 10.2   Schematic of hybrid gamma camera imaging system. The arrangement of the mirror and detectors creates identical field of views for optical and gamma images at any imaging distance.

at the camera face. The tungsten collimator (Ultimate Metals, Bedfordshire, UK) contains a 1.0 mm diameter knife-edge pinhole with an acceptance angle of 60°, with the pinhole centre at 10 mm from the detector. The calculated on-axis sensitivity to 140 keV gamma photons is 1.4×10^{-4} at the camera face and the calculated spatial resolution is 3.5 mm at a distance of 60 mm from the collimator.

A 1.0 mm thick first surface mirror (UQG Optics Ltd, Cambridge, UK) is positioned in front of the pinhole at a 45° angle (see Figure 10.2). Gamma photons pass through the mirror with minimal absorption and scatter whereas optical photons are reflected towards a high-resolution optical camera (IDS Imaging Development Systems, Obersulm, Germany) outside of the direct line of sight of the pinhole. The positioning of the optical camera is such that the ratio of the magnification of the optical and gamma image is independent of imaging distance.

10.3.2 Imaging Software

A full gamma image consists of many individual frames, with 10 frames acquired per second of image accumulation. Individual frames are

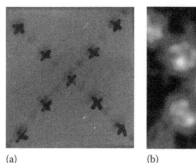
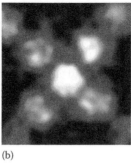
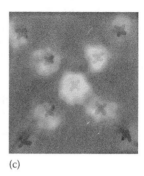

(a)   (b)   (c)

FIGURE 10.3 **(See colour insert.)** Illustration of single- and dual-modality hybrid gamma camera images using a bespoke alignment phantom. (a) Optical image, crosses indicate wells containing activity. (b) Gamma image showing the expected nine areas of activity. (c) Hybrid image showing good alignment between the modalities. The phantom contained 13 MBq of ^{99m}Tc and was imaged from a distance of 11 cm. Exposure time was 3.5 min.

analysed using a 'blob detection' algorithm with automatic scale selection (Lees et al. 2011). This analysis determines the position of each gamma interaction within the scintillator along with the energy of the detected photon.

The accumulated gamma and optical images may be displayed separately or combined into a single fused image. Figure 10.3 illustrates these options using a phantom designed at the University of Leicester to test optical and gamma image alignment. Colour tables and contrast are adjustable within the acquisition and display software to aid interpretation.

10.3.3 Performance Characteristics

The HGC system has been fully characterised using the protocols outlined in Chapter 9. Table 10.1 contains a summary of these results. Where camera performance is known to vary with imaging distance, the relationship between the parameter and d – defined as the distance from the camera face in mm – has been provided. Further details are provided in Bugby et al. (2014).

10.4 PHANTOM IMAGING

A range of phantoms have been used in experimental studies to fully understand the performance of the HGC system. This section discusses

TABLE 10.1   Performance Characteristics for Hybrid Gamma
Camera System with a 1.0 mm Diameter Pinhole Fitted and No Energy
Windowing Applied

FOV (mm)	At camera face	20×20
	At d	$0.8 \times (d + 25)$
Intrinsic spatial resolution (mm)	FWHM	0.32
	FWTM	0.58
System spatial resolution (mm)	FWHM at camera face	3.3
	FWTM at camera face	6.2
	FWHM at d	$3.3 + 0.06d$
	FWTM at d	$6.2 + 0.12d$
Intrinsic sensitivity	At 141 keV	30%
System sensitivity (cps/MBq)	At camera face	12.5
	At 100 mm	0.064
Uniformity	Coefficient of variation	31%
	Differential uniformity	31%
Energy resolution	At 141 keV	130%

Note: The parameter d is the imaging distance from the camera face in mm,
through scattering material.

a number of phantom designs and provides examples of the images that
have been obtained.

10.4.1 Thyroid Phantom

The Picker thyroid phantom (Part #3602 [Marconi Medical Systems {for-
merly Picker}, Cleveland, OH]) provides a simulation for thyroid imaging. The phantom is filled with an appropriate liquid radionuclide source
and features a number of structures which may be encountered in thyroid
imaging (see Figure 10.4).

The wide availability of the Picker phantom makes it a good candidate
for performance comparisons between systems for measurements such
as contrast-to-noise ratio (Bugby et al. 2016). Figure 10.5 shows a time
sequence for a 15 min accumulation used to investigate imaging time and
detectability.

As expected, the clarity of the phantom image improves with increased
acquisition time and count rate. The optimal acquisition time for clinical
imaging will vary depending on the application and the activity at the
site under investigation. In clinical use the amount of activity used and
acquisition parameters will rest with the clinician and the nuclear medicine team.

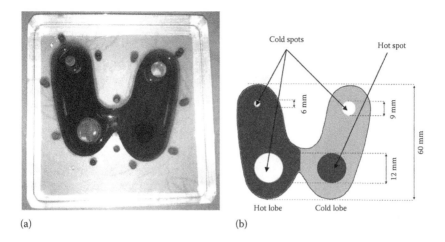

(a) (b)

FIGURE 10.4 Photograph (a) and schematic (b) of Picker thyroid phantom. White indicates a cold spot, light grey half depth and dark grey full depth (18.4 mm). Total phantom size is approximately 60 mm × 60 mm.

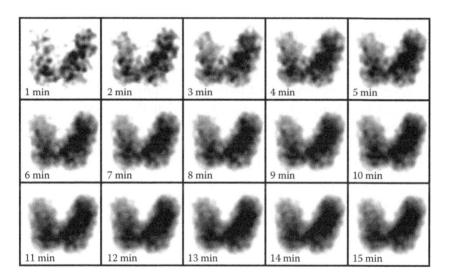

FIGURE 10.5 A cumulative time series of thyroid phantom images. Each frame shows an increment in integration time of 1 min. The phantom contained 15 MBq of ^{99m}Tc in 60 mL and was imaged from a distance of 10.5 cm.

10.4.2 Lymphatic Drainage Phantoms

A lymphoscintigraphic phantom has been designed using widely available components (Alqahtani et al. 2015). The lymphatic vessels have been simulated with a cannula, a sentinel lymph node with a small well source

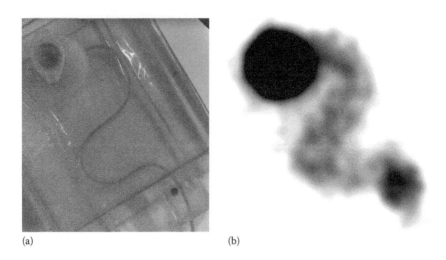

(a)　　　　　　　　　　　　　(b)

FIGURE 10.6　(a) Optical image of lymphatic simulation with a large injection site (top left) and small sentinel node (bottom right). (b) Gamma image of lymphatic simulation. The injection site contained 50 MBq of ^{99m}Tc in 1.0 mL, the lymphatic vessel 3 MBq in 0.3 mL and the sentinel node 0.6 MBq in 0.06 mL. The phantom was imaged at a distance of 10.5 cm and exposure time was 2 min.

and the bright injection site with an Eppendorf tube (see Figure 10.6) with the activities of each component tailored to the procedure being investigated. Layers of Perspex may be used to simulate the scattering effect of body tissues.

This phantom is useful for simulating sentinel lymph node procedures where the sentinel node is in close proximity to the higher activity at the injection site. In Figure 10.6, the hot injection site, simulated lymph vessel and low activity sentinel node can all be observed.

This phantom may be used to investigate detectability for a range of injection site to sentinel node activity ratios and separation distances.

10.4.3 Contrast Phantoms

A bespoke contrast phantom has been designed and fabricated at the University of Leicester to investigate contrast-to-noise ratio. This phantom does not aim to simulate a clinical procedure but to allow efficient and replicable measurements to be taken.

The phantom consists of a number of well sources with diameters ranging 10–2.5 mm. These wells are duplicated for simultaneous testing of different source or background activity. A separate plate contains four square

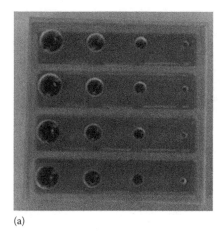

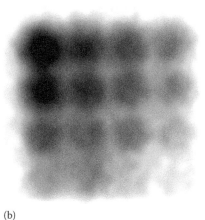

(a)   (b)

FIGURE 10.7 (a) Optical image of contrast phantom showing circular sources of varying diameter and rectangular wells for the simulation of background activity. (b) Gamma image of contrast phantom. The sources were filled with ^{99m}Tc with an activity of 3, 1.8, 1.0 and 0.2 MBq (left to right) in the top row with activities halving in each subsequent row. Background activity concentrations were one tenth that of the nodes in each row. The phantom was imaged at a distance of 12 cm and exposure time was 13 min; the image has been square root compressed to improve the dynamic range.

wells for background activity. The background plate may be orientated to give different combinations of well size, activity and background (see Figure 10.7).

10.5 PATIENT IMAGING

As part of the evaluation of the HGC system, patients undergoing clinical investigations at the Nuclear Medicine Department at Queen's Medical Centre, Nottingham University Hospitals NHS Trust were imaged with the HGC following their standard clinical test. This exploratory study was carried out following ethical and local Research and Innovation Directorate approval.

This section presents a couple of clinical examples of patient images using the prototype camera.

10.5.1 Thyroid Imaging

A 49-year-old male patient undergoing thyroid scintigraphy was administered a standard dose of 18.5 MBq ^{123}I-NaI (159 keV) intravenously

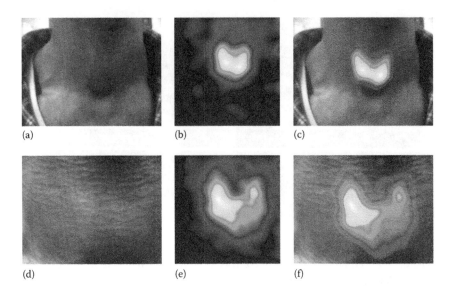

(a)   (b)   (c)

(d)   (e)   (f)

FIGURE 10.8 **(See colour insert.)** (a–c) Optical, gamma and combined anterior images of the neck at ~17 cm (300 s image acquisition) starting at 149 min post injection. (d–f) Optical, gamma and combined anterior images of the neck at ~8 cm (300 s image acquisition) starting at 142 min post injection. Gamma images are shown with a 5 pixel width smoothing filter applied.

140 min prior to imaging. Figure 10.8 shows two sets of images from the prototype camera at a distance of 170 mm (top) and 80 mm (bottom).

The thyroid images show greater activity in the right lobe with less activity present in the left lobe (i.e. on the RHS of the images in Figure 10.8). A region of higher uptake can be seen at the top of the left node. This pattern of uptake was confirmed by imaging (not shown) using a standard large field of view (LFOV) gamma camera. Quantification from the LFOV image indicated that 19.5% of the administered activity was taken up by the thyroid. This gave an estimation of thyroid activity of 3.2 MBq at the time of imaging, suggesting an HGC sensitivity of around 1.6 cps/MBq at 80 mm and 159 keV.

10.5.2 Lacrimal Drainage

A 62-year-old female patient undergoing a radionuclide lacrimal drainage scan was administered a 1 MBq of ^{99m}Tc-pertechnate in saline solution to each eye prior to patient imaging. Figure 10.9 shows the HGC prototype hybrid images alongside a LFOV system gamma image of both eyes.

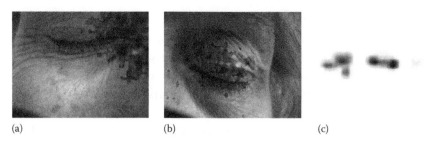

(a)   (b)   (c)

FIGURE 10.9 **(See colour insert.)** Hybrid lacrimal drainage scan from the HGC system (a, b) and conventional large field of view (LFOV) lacrimal drainage scan. Hybrid gamma camera (HGC) images taken 1 h 35 min (a) and 1 h 42 min (b) after administration and the conventional LFOV lacrimal drainage scan (c). HGC imaging was at a distance of 7 cm with an exposure time of 5 min.

The hybrid and LFOV images are in general agreement. In this case, imaging was carried out with a version of the HGC using a thinner scintillator (600 μm thick scintillation crystal) which is less sensitive than that used in the other studies described in the rest of this chapter (Lees et al. 2011, Bugby et al. 2014).

10.6 FUTURE DIRECTIONS

Further development is ongoing at the Universities of Leicester and Nottingham to extend the HGC concept.

Stereoscopic imaging with depth estimation of the gamma-emitting source could be achieved by analysing images from two identical cameras with a known separation between them. Recently, there has been a huge drive to produce 'glasses free' 3D image displays and such devices, when fully realised, would obviously be ideal for stereoimaging. In addition to providing a qualitative 3D image, a stereoscopic configuration (producing two optical and two gamma images in total) allows the calculation of object distances or gamma source depths below a surface.

Recently, there have been a number of studies that have shown that hybrid fluorescent-radioactive tracers, such as ICG-^{99m}Tc-nanocolloid, improve the identification of sentinel lymph nodes compared with the standard technique using blue dye (Brouwer et al. 2012, van den Berg et al. 2012). However, this new approach still relies on invasive non-imaging gamma detectors to locate the nodes.

We are aiming to develop a new hybrid camera system that will be able to image both the fluorescent and the gamma tracers simultaneously, offering in situ real-time imaging of the target nodes whilst visualising the surrounding healthy tissue. Such a system will enable multiplexing of fluorescent and gamma tracers during the same surgical procedure. Multiplexing will allow the surgeon to use a combination of tracers to visualise different physiological and cellular markers, improving the overall sensitivity and specificity of detection.

To date, there is no commercial portable system offering a hybrid combination of fluorescence and gamma imaging in a single fused image. Development of a combined fluorescence and gamma system is expected to have a significant benefit for diagnosis and treatment, for example, in the case of head and neck cancers and in the assessment of tumour margins.

10.7 SUMMARY AND CONCLUSIONS

Intraoperative imaging is used widely in medicine, and the introduction of gamma camera imaging has the potential to further improve surgical outcomes. For example, in sentinel lymph node biopsy, assisting in the identification of patients for complete lymph node dissection (Koops et al. 1999, Salvador et al. 2007) and in particular, determining whether the disease has spread from the primary tumour to the lymphatic system. This procedure is becoming standard using gamma probes and is carried out over two million times annually.

The clinical advantages of intraoperative imaging, fusing the gamma image to the optical image are clear and should be especially beneficial for use during surgical procedures. The HGC system may have many practical benefits for a number of clinical procedures including diagnosis, surgical investigation and small organ imaging.

We anticipate that by providing the surgeon with a fused image during an operation can increase the confidence in localisation of sites of uptake, reducing the procedure time and providing a better diagnosis and patient outcome.

The HGC has been shown to be capable of high-resolution gamma imaging combined with optical imaging to produce fused images offering improvements in a number of nuclear medicine applications. The camera has been used to produce dual-modality images from phantom simulations and patients undergoing clinical examinations. In particular, HGC is able to detect and image ^{123}I uptake in a patient thyroid with an acquisition time comparable to that for LFOV clinical studies (Bugby et al. 2016).

HGC offers combined optical and gamma imaging that will facilitate new techniques to address both existing and emerging healthcare needs, especially point-of-care and intraoperative imaging. The development of these devices will offer exciting new horizons for optical and nuclear imaging.

REFERENCES

Alqahtani, M.S., J.E. Lees, S.L. Bugby, L.K. Jambi and A.C. Perkins. 2015. Lymphoscintigraphic imaging study for quantitative evaluation of a small field of view (SFOV) gamma camera. *Journal of Instrumentation*, 10(07): P07011.

Beyer, T., L.S. Freudenberg, J. Czernin and D.W. Townsend. 2011. The future of hybrid imaging – Part 3: PET/MR, small-animal imaging and beyond. *Insights into Imaging*, 2(3):235–246.

Brouwer, O.R., T. Buckle, L. Vermeeren et al. 2012. Comparing the hybrid fluorescent–radioactive tracer indocyanine green–^{99m}Tc-nanocolloid with ^{99m}Tc-nanocolloid for sentinel node identification: A validation study using lymphoscintigraphy and SPECT/CT. *Journal of Nuclear Medicine*, 53(7): 1034–1040.

Bugby, S.L., J.E. Lees, B.S. Bhatia and A.C. Perkins. 2014. Characterisation of a high resolution small field of view portable gamma camera. *Physica Medica*, 30(3):331–339.

Bugby, S.L., J.E. Lees, A.H. Ng, M.S. Alqahtani and A.C. Perkins. 2016. Investigation of an SFOV hybrid gamma camera for thyroid imaging. *Physica Medica*, 32(1):290–296.

Cherry, S.R. September 2009. Multimodality imaging: Beyond PET/CT and SPECT/CT. *Seminars in Nuclear Medicine*, 39(5):348–353.

Duch, J. 2011. Portable gamma cameras: The real value of an additional view in the operating theatre. *European Journal of Nuclear Medicine and Molecular Imaging*, 38(4):633–635.

Koops, H.S., M.E. Doting, J. de Vries et al. 1999. Sentinel node biopsy as a surgical staging method for solid cancers. *Radiotherapy and Oncology*, 51(1):1–7.

Lees, J.E., D.J. Bassford, O.E. Blake, P.E. Blackshaw and A.C. Perkins. 2011. A high resolution Small Field Of View (SFOV) gamma camera: A columnar scintillator coated CCD imager for medical applications. *Journal of Instrumentation*, 6(12):C12033.

Morris, P. and A. Perkins. 2012. Diagnostic imaging. *The Lancet*, 379:1525–1533.

Olcott, P., G. Pratx, D. Johnson, E. Mittra, R. Niederkohr and C.S. Levin. 2014. Clinical evaluation of a novel intraoperative handheld gamma camera for sentinel lymph node biopsy. *Physica Medica*, 30(3):340–345.

Perkins, A.C. and J.G. Hardy. 1996. Intra-operative nuclear medicine in surgical practice. *Nuclear Medicine Communications*, 17(12):1006–1015.

Salvador, S., V. Bekaert, C. Mathelin, J.L. Guyonnet and D. Huss. 2007. An operative gamma camera for sentinel lymph node procedure in case of breast cancer. *Journal of Instrumentation*, 2(07):P07003.

Tsuchimochi, M. and K. Hayama. 2013. Intraoperative gamma cameras for radioguided surgery: Technical characteristics, performance parameters, and clinical applications. *Physica Medica*, 29(2):126–138.

van den Berg, N.S., O.R. Brouwer, W.M.C. Klop et al. 2012. Concomitant radio- and fluorescence-guided sentinel lymph node biopsy in squamous cell carcinoma of the oral cavity using ICG-^{99m}Tc-nanocolloid. *European Journal of Nuclear Medicine and Molecular Imaging*, 39(7):1128–1136.

Visioni, A. and Kim, J. 2011. Positron emission tomography for benign and malignant disease. *Surgical Clinics of North America*, 91(1):249–266.

Zaidi, H. and R. Prasad. 2009. Advances in multimodality molecular imaging. *Journal of Medical Physics*, 34(3):122.

Index

Printed and bound by CPI Group (UK) Ltd, Croydon, CR0 4YY

01/11/2024

01782618-0003